Feminist Technosciences

Rebecca Herzig and Banu Subramaniam, Series Editors

Figuring the Population Bomb

GENDER and DEMOGRAPHY in the MID-TWENTIETH CENTURY

CAROLE R. McCANN

UNIVERSITY OF WASHINGTON PRESS
Seattle and London

Earlier versions of chapters 2 and 5 originally appeared in Carole McCann, "Malthusian Men and Demographic Transitions: A Case Study of Hegemonic Masculinity in Mid-Twentieth Century Population Theory," special issue, *Frontiers* 30, no. 1 (May 2009): 142–171. They are included here with the permission of the University of Nebraska Press.

Printed and bound in the United States of America
Design by Thomas Eykemans
Composed in Chaparral, typeface designed by Carol Twombly
21 20 19 18 17 5 4 3 2 1

UNIVERSITY OF WASHINGTON PRESS
www.washington.edu/uwpress

LIBRARY OF CONGRESS CATALOGING-IN-PUBLICATION DATA
Names: McCann, Carole R. (Carole Ruth), 1955– author.
Title: Figuring the population bomb : gender and demography in the mid-twentieth century / Carole McCann.
Description: 1st Edition. | Seattle : University of Washington Press, 2016. | Series: Feminist technosciences | Includes bibliographical references and index.
Identifiers: LCCN 2016014273| ISBN 9780295999098 (hardcover : alk. paper) | ISBN 9780295999104 (pbk. : alk. paper)
Subjects: LCSH: Population. | Population policy. | Population—Social Aspects. | Birth control—History—20th century. | Birth control—Political aspects.
Classification: LCC HB851 .M296 2016 | DDC 304.6—dc23
LC record available at https://lccn.loc.gov/2016014273

The paper used in this publication is acid-free and meets the minimum requirements of American National Standard for Information Sciences—Permanence of Paper for Printed Library Materials, ANSI Z39.48–1984. ∞

For Mel Holden

Contents

Acknowledgments

I have many people to thank for their support and encouragement during the years I worked on this project. First and foremost I wish to acknowledge the valuable support provided by two faculty fellowships from the University of Maryland, Baltimore County: the Provost Research Fellowship, which supported a yearlong sabbatical at the beginning of this project, and the Dresher Center of the Humanities 2014 Faculty Fellowship, which arrived at a crucial juncture near the end of the project. The Dresher Fellowship, and the well-lit office it provided, gave me the time and space to complete the research for chapter 6. I also extend my thanks to the Rockefeller Archive Center for access to and permission to cite records of the Population Council.

I am grateful to the vital intellectual community in and around Gender and Women's Studies at UMBC. Their continued commitment to our collective intellectual enterprise makes my scholarly work possible. Thanks to Rebecca Adelman, Jessica Berman, Bev Bickel, Dawn Biehler, Gloria Chuku, Amy Froide, Marjoleine Kars, Christine Mallinson, Susan McCully, Patrice McDermott, Susan McDonough, Michelle Scott, Orianne Smith, Lisa Pace Vetter, Shelly Wiechelt, and especially the core GWST faculty: Amy Bhatt, Kate Drabinski, and Viviana MacManus. I also extend my special thanks to two dear friends: Tulay Adali, with whom I have shared many adventures and hours of thoughtful conversation; and my collaborator, Seung-kyung Kim, with whom I have worked closely on the several editions of *Feminist Theory Reader.* Our ongoing conversations about life, theory, research, and the joys of university life have been a continued source of encouragement and good humor.

The intellectual exchange with students is another important source of creative energy. There is nothing I enjoy more than sharing new ideas with them. I also want to acknowledge those who have supported research for this book. They include Rachel Carter, Susie Hinz, Lindsay Loeper, Autumn Reed, Melissa Smith, Arlene Barrow, and especially Emek Ergun, who worked with me on this project for many years and cheerfully proofread more versions of these chapters than anyone should have had to.

Every project has pivotal moments. Mine came after this project had been incubating for several years, when I was lucky enough to see the call for the Feminist Epistemology, Methodology, Metaphysics and Science Studies Second Conference (FEMMSS 2). The colleagues I met through FEMMSS and the feedback on my work at that and at FEMMS 3, 4, and 5 enabled me to move forward on what had been a stalled project. My particular thanks go to Drucilla Barker, Suzanne Bergeron, Nancy Campbell, LeeRay Costa, Mary Margaret Fonow, Aya Homei, Patti Lather, and Laura Parisi for their comments on papers at conferences and in reviews. I also am particularly indebted to the small group of feminist science studies scholars who meet in venues such as FEMMSS and the National Women's Studies Association Science Studies Task Force. Their continued insight and friendship have been invaluable to me. They include Rajani Bhatia, Virginia Eubanks, Laura Foster, Sarah Giordano, Kasi Jackson, Clare Jen, Jane Lehr, Deboleena Roy, Chikako Takeshita, Angela Willey, and Mary Wyer. I am particularly grateful to Banu Subramaniam for encouraging me to submit this work to the University of Washington Press. I am honored to have this work included in their new feminist technosciences series. My thanks also to Rebecca Herzig, Larin McLaughlin, Jacqueline Volin, and everyone at the University of Washington Press for their help and encouragement throughout the publication process.

Finally, I owe an incalculable debt to my partner and best friend, Mel Holden. When we first met, he said he was looking for someone with whom he could share what joys might come and what sorrows might befall. We have done both with love and honesty. But also, not being an academic himself, he has learned to accept with grace the challenges of living with someone engaged in the rigors of scholarly writing while running an academic department. In addition to his proofreading abilities and technical assistance with the figures presented here, I am grateful for his steadfast love and support. It makes all things possible.

Abbreviations

ABCL	American Birth Control League
ICPD	International Conference on Population and Development
IPPF	International Planned Parenthood Federation
IUD	Intrauterine device
IUSSP	International Union for the Scientific Study of Population (1947–present)
IUSSPP	International Union for the Scientific Study of Population Problems (1928–1947)
IWHC	International Women's Health Coalition
KAP	Knowledge, attitudes, and practices
MMF	Milbank Memorial Fund
MSP-LC	Margaret Sanger Papers, Library of Congress
NCMH	National Committee on Maternal Health
NGO	Nongovernmental organization
OPR	Office of Population Research, Princeton University
PAA	Population Association of America
PPFA	Planned Parenthood Federation of America
RAC	Rockefeller Archive Center, Sleepy Hollow, New York
RAC-PC	Rockefeller Archive Center, Population Council Collection

Figuring the Population Bomb

1

Matters of Vital Importance

Demography and the Mid-Twentieth-Century Population Imaginary

> Discourses are not just "words"; they are material-semiotic practices through which objects of attention and knowing subjects are both constituted.
>
> —DONNA HARAWAY

> Closer inspection reveals that the numbers work most effectively in a world they have collaborated in creating.
>
> —THEODORE PORTER

IN MAY 1968, THE SIERRA CLUB AND BALLANTINE BOOKS PUBlished *The Population Bomb,* by Paul Ehrlich. The book's doomsday scenario was announced on the cover. Below the title, in a linotype-style font highlighted in bright yellow, was the warning "WHILE YOU ARE READING THESE WORDS FOUR PEOPLE WILL HAVE DIED FROM STARVATION. MOST OF THEM CHILDREN." In the lower left corner, an illustration of a cannonball-shape bomb appeared in a box above the words "The population bomb keeps ticking."[1] The book opens with Ehrlich's declaration that he has long understood the population explosion intellectually, but "came to understand it emotionally one hot stinking night in Delhi": "The streets seemed alive with people. People eating, people washing, people sleeping. People visiting, arguing, and screaming. . . . People, people, people, people. As we moved slowly through the mob, hand horn squawking, the dust, noise, heat, and cooking fires gave the scene a hellish aspect. . . . since that night I've known the *feel* of overpopulation."[2] Ehrlich then presents the academic case for overpopulation, noting that "no matter how you slice it, population is a numbers game."[3] The problem: after millennia of slow increases in human populations, an accelerating rate of growth meant that the "doubling time"[4]—the amount of time needed for world population to double in size—had decreased suddenly and dramatically. Moreover, as exemplified by his vignette of India, populations were not growing "uniformly over the face of the Earth" but appeared to be concentrated in "underdeveloped countries."[5] The solution: conscious regulation of human numbers through population control (voluntary contraception for the self-disciplined, mandatory contraception for those who are not), which, with vigilance, might defuse the population bomb in time.[6]

Ehrlich was not the first to use bomb imagery. Demographers had suggested it in the first accounts of sudden, rapid growth in the mid-1940s. *The Population Bomb* also was the title of Hugh Moore's 1954 self-published pamphlet, which he regularly updated for more than a decade. During the midcentury years when the works of Moore and Ehrlich circulated, the phrases *population bomb* and the more common *population explosion* became firmly established as fact in the social imaginary.[7] Historians and demographers generally dismiss Moore and Ehrlich as ethnocentric alarmists who overstated the crisis and underestimated the power of technical progress to accommodate growth. But the population figures that grounded their arguments were not—and are not—questioned.[8] Those figures—both the numbers and their explosiveness—were the products of US demography, and their impact endures.[9]

A relatively young social science, demography blossomed in the mid-twentieth century as part of the US-dominated development paradigm. With boundless confidence in science, the post–World War II generation of US social scientists studied and prescribed technical solutions to the problems of world "hunger" and "want," which were then perceived as testing American international leadership.[10] Drawing on Malthusian discourse, US demographers identified rapid population growth as an underlying cause of poverty and offered population control as the solution. They staked their epistemic authority on innovative statistical techniques and organizational sites for measuring population dynamics.[11] In response to what they claimed was the "chaotic state" of population knowledge after the war, they worked through the United Nations to standardize global data gathering.[12] Through disciplinary powerhouses such as the Princeton University Office of Population Research (OPR) and the Population Council, an NGO founded by John D. Rockefeller III, US demographers dominated the production of knowledge about current population patterns and projections of future trends. By 1968, they had helped organize and fund half a dozen or more academic departments and research centers in the United States, Asia, and Latin America.[13] Together these institutions trained a multinational cadre of population scientists who staffed statistical bureaus in emerging nations of the Global South. US demographers also provided technical assistance to family planning programs around the world. This network of regional, national, international, and nongovernmental organizations provided the institutional and epistemic grounding for the official statistics by which the new problems of rapid population growth and its control were understood and worried over.[14] The confidence with which demographic figures were quoted in the press, in related scientific disciplines, in public policy debates, and at cocktail parties shows how effectively demographic statistical reasoning assembled the credibility of mid-twentieth-century population facts.[15] The extent to which population facts moved individuals, NGOs, and nations to ever more urgent interventions into procreative processes demonstrates how much demographic numbers mattered.

Much attention has been paid to the most visible effect of the mid-twentieth-century population crisis: population control through state-sponsored family planning. In particular, transnational feminist scholarship and activism have excavated the racial, class-based, and imperialist gender biases underlying coercive population control practices across the globe.[16] This invaluable work has been critical to political efforts to end such practices. Yet the tight focus on population control practices has limited our understanding of the politics of gender, race, class, and "coloniality" inscribed in population knowledge.[17] That is, the history of the demographic facts and figures that fueled the population panic and animated population control has been largely ignored.[18] The production of those facts—the disciplinary milieu, institutional infrastructure, affective economies, and inferential modalities—is obscured by the solidity of the numbers.[19] Even in demographic texts, details about the construction of the numeric figures are relegated to appendices, which are generally decipherable only by experts.[20] But as Susan Greenhalgh notes in her important study of population governance in contemporary China, it is "in the making" of "numeric inscriptions—those mundane tables, figures, charts, and equations—. . . that population scientists . . . do some of their most important yet least studied work."[21]

This book traces the social-epistemic genealogy of mid-twentieth-century demographic knowledge.[22] It shows how the history of demographic facts and figures is crucial to a fuller understanding of the gendered geopolitics that authorized the conscious limitation of fertility as a governing technology and a normative requirement of modern citizenship globally. Beginning with the observation that scientific knowledge is produced in specific times and places, and inscribes those circumstances within it, the analysis investigates the social, affective, and epistemic contours of the context in which demographers "discovered" the population crisis.[23] The following chapters investigate the complex and shifting relationships between the social worlds of demography, its publics, and its rivals, highlighting the tensions and accommodations that have delimited the discipline's boundaries.[24] Through close readings of foundational texts, disciplinary histories, participant reminiscences, and archival records, the analysis elucidates the grounds of the "shared mathematical culture," "bureaucratic organization," and "intertextual hierarchy" that assembled the facts of the midcentury population crisis and gave them epistemic and affective authority.[25] In particular, the analysis excavates the gendered geopolitics of the social networks, cultural anxieties, and discursive modalities that configured the "knowing subjects," "rituals of truth," "domain of objects," and "global designs" that held mid-twentieth-century population figures together.[26] It demonstrates that the population crisis problematization was not merely a gloss on the data or a misapprehension of the facts. Rather, it was a constitutive element of the historically situated "knowledge culture" in which demographic questions were asked and the numbers were calculated.[27] The analysis demonstrates that demographic theory and measurement practices constructed "procedure[s] of reasoning"[28] about fertility, mortality, growth, modernization, and economic development that ignited the population crisis and moved nations to act.[29] Demographic knowledge configured a mathematical panic that pathologized population growth in the Global South and intensified the scrutiny of and interventions into women's reproductive lives.

Hegemonic Masculinity and Manly States

The analysis brings into focus the gendered grammar of coloniality inscribed in demographic facts and illuminates the cultural work they performed within the interconnected cognitive frameworks and affective alignments of mid-twentieth-century hegemonic masculinity.[30] It uses a transnational feminist science studies lens that takes account of the temporal, spatial, and affective specificity of the configuration of scientific knowledges and practices. It recognizes the multiple and heterogeneous ways that the effects of science play out in different times and places, while also recognizing the encumbrances of the gendered, racialized coloniality of knowledge shaping those histories. Moreover, rejecting the positivist splitting of knowledge and emotion, this analysis recognizes that knowledge cultures, as socially and historically situated sites of human endeavor, are always imbued with configurations of normative sentiments that bind ideas and affects together in politically inflected tangles.

One component of the analytic lens used here draws on Sara Ahmed's explication of how "accumulation of affective value shapes the surfaces of bodies and worlds." Her work on the cultural politics of emotion illustrates that as affects circulate in social historical worlds, repetitive associations between some signs, objects, and figures become "sticky." That stickiness binds them tightly together and immobilizes them.[31] Using the tools of cultural analysis, feminist interrogation of such immobilized signs opens space to "imagine otherwise." The signs of overpopulation—excessive fertility, poverty—and the bodies of Third World women compose one such sticky figuration in the flow between signs and bodies in and through population discourse. The analysis brings into focus another less obvious but vitally important sticky figure constructed in this discourse, the Malthusian man, who binds mathematics and masculinity into the authoritative subject of population knowledge. By tracing the repetitive association of these signs and bodies in specific sites of their construction, the analysis exhumes the affects, the structure of feeling that has organized population knowledge and, thereby, opens space to "learn" the "lessons" of the population explosion "otherwise."[32]

The concept of hegemonic masculinities, developed by sociologist Raewyn Connell, is a second particularly useful tool for excavating the complicated and multiple assemblages of gender, race, and coloniality in demographic figures.[33] It locates gender in the social processes that organize "the reproductive arena," both in "configurations of practice" and "discursive constructions." From this point of view, "masculinity and femininity are produced together in a process that constitutes the gender order." Moreover, binary masculinities and femininities are conceptualized as both biosocial places in gender relations and configurations of affects and practices that engage those places.[34] Following Gramsci's conceptualization, hegemonic masculinity theory assumes that "active struggle for dominance" is "implicit" in gender regimes. Gender struggles involve multifaceted contestations over both the boundaries between and the hierarchies within gender categories. Hegemonic masculinity is thus "the configuration of gender practice which embodies the currently accepted answer to the problem of the legitimacy of patriarchy" as both subordinations of women/femininity and hierarchies of men/masculinity. That is, "hegemonic masculinity is hegemonic [in] so far as it embodies a successful strategy in relation to women" that "guarantees" men's dominance and women's subordination. But in addition to settling relations between genders, hegemonic masculinity implies "the ability to impose a particular definition on other kinds of masculinity."[35] Historically and geographically situated gender regimes are thus the products of always tense, contingent relations between and within gender groups. Moreover, gender regimes, configured in local milieus and moments, are never stable. Contested by women, masculine rivals, and masculine subordinates, settled gender relations are subject to recurring crises.[36]

Configurations of gender become hegemonic to the extent that they are invested in cultural discourses and institutional practices that allocate power, authority, legitimacy, and emotionality in relation to gender exemplars. These exemplary figures embody narratives of masculine mastery that explain why things "are as they are" and "what the future may portend."[37] They provide a way for some men to position themselves through discursive practices that effectively "ward off anxiety and avoid feelings of powerlessness."[38] Such exemplary masculinities are configured in contrast to subordinated, marginalized masculinities, which are composed by projecting "currently unwanted characteristics onto subordinate groups, branded as pathological."[39] At the same time, exemplary masculinity is configured in relation to emphasized (subordinate) femininity, coupled to it in the heteronormative gender binary. That is, characterized by distance, embrace, and domination, hegemonic masculinity dreams up women to fall in love with. Emphasized femininities are distinguished by their lack of masculinity, but they are oriented toward the interests and desires of masculinity and emphasize feminine compliance with subordination as desired objects.[40] Of course, femininity is scaled as well in terms of deviation from the idealized women coupled to exemplary men. This theorization of femininity is capacious enough to encompass women as active participants in the determination of relations between and within genders, while also illuminating how they are disadvantaged in those struggles with regard to masculinities.

Connell has used the concept of hegemonic masculinity to analyze the historicity of global gender regimes, which, she argues, are the products of gender power relations configured in specific times and places, but that are mobile. Those power relations are themselves imbricated with race, class, and colonial processes. Therefore, gender has been a central configuration of practice in European empires. That is, the "imperial social order" configured "a scale of masculinities" as it "created a scale of communities and races."[41] In the mid-twentieth century, "US imperialism-without-colonies" drew upon and rearticulated earlier scales of coloniality in the modernization and development regime.[42] US demography, which was heir to Thomas Malthus's imperial-age principle of population, would become a constitutive social science of the modernization regime.[43] The chapters herein show that mid-twentieth-century demographic figures were composed in the sum of the densely particular technical decisions taken in the light of grand social theory and unsettling historical circumstances; thus they were shaped by and in turn reinforced that era's formation of global hegemonic masculinity. This gendered geopolitics of population knowledge continues to haunt feminist efforts to secure reproductive justice.[44] Understanding demography's complex gendered coloniality is especially important for those who, as downstream consumers, must continue to use demographic knowledge products in pursuit of social justice.

The analysis here pushes hegemonic masculinity theory in new directions by focusing attention on the tangled interconnections between gender binaries and gender hierarchies in population knowledge and politics.

Noting that classification systems "reveal as much about the classifiers, their prejudices and anxieties about their own identities, as they do about the classified,"[45] the analysis elucidates how demographic discourses co-produced privileged knowers and objects of epistemic attention.[46] That is, it shows how demographic theory and measurement practices configured white First World men as their subjects as it configured poor Third World women's fertility as its analytic object. By turning the focus back on the makers of knowledge, this analysis illuminates the masculinities staked within reproductive politics. In this sense, it responds to the call from reproductive health advocates to bring men back in.[47] Further, it examines how demographic theory and measurements configured a stratified scale of femininity in which Euro-American women represented modern, reproductively responsible femininity, and women of the Global South represented a pathological "reproductive performance."[48] Finally, it investigates how these feminine fertility figures became fetishized tokens for ranking groups of men, nations, and forms of masculinity on the geopolitical scales of global modernity.[49] It thereby elucidates how the intersecting contests within and between genders in the reproductive arena shaped our common understanding of population dynamics and its differing effects on the reproductive lives of women worldwide.

Discovery and Travels of the Population Crisis

Demographers articulated the population crisis problematization as world leaders faced the aftermath of the second global war. Responding to a number of overlapping crises, Harry Truman announced a new "fair deal" in American foreign policy in his 1949 inaugural address. Faced with the collapse of European colonialism, the rise of Communism, and the "discovery" of "world poverty," Truman authorized extensive foreign aid to provide modern market- and science-based technologies to "underdeveloped areas" of the world as the best means for replicating "the features that characterized the 'advanced' societies . . .'"[50] The hope was that this "fair deal" would reduce poverty and the risk of Communist revolution. Along with economics, sociology, and international relations, demography participated in the social science knowledge culture of the modernization and development regime that configured the Third World in discourse and practice.[51] While the specific trajectories of modernization programs and their effects developed differently in regions depending on their histories of colonialism, anticolonialism, and perceived importance to US Cold War strategic interests, replication of Western development processes and forms became a benchmark of "democratic" modernity. And as Arturo Escobar has noted, "The policies and programs that originated from this vast field of knowledge inevitably carried with them strong normalizing components."[52] The demographically produced population crisis instantiated reduced fertility rates among those norms of modernity.[53] At the same time, the population crisis narrative positioned the discipline of demography as a necessary component of the technocratic social sciences and a requirement of modern citizenship.

Political and epistemic realignments embodied in modernization theory involved changes in gender hierarchies and binaries as well. The modernization project for the Third World embodied the new configurations of "colonial difference."[54] In the wake of Nazi genocide, the collapse of European colonialism, and growing civil rights movements worldwide, biological theories of race differences were increasingly discredited. Cultural theories of race, which Étienne Balibar has called "differentialist culturalism," located racial difference in cultural heritage instead of genetic heredity. Social scientific discourses rearticulated associations between "states, peoples, and cultures on a world scale," situating their cultural inheritances, particularly family forms and religion, as the site of racial difference production.[55] But, as with all racial philosophies, differentialist culture theory "ma[de] visible the invisible cause of the fate of societies and peoples." Moreover, the new cultural discourses "le[d] to the same old acts" and continued to "legitimate policies of exclusion."[56] They also reaffirmed for the West the superiority of its culture, while the rest were represented as variously weighed down by their differences.

The focus on kinship systems and religion firmly connected the scale of cultures to the gender binary of the reproductive arena, which was itself subject to renegotiation. Social theorists such as Talcott Parsons, who identified families organized by conventional (white, middle-class) sex roles as essential to a stable society, contributed to the reconsolidation of modern masculinist gender relations.[57] While Parsonian sex-role theory involved concessions to women's demands for greater public roles—the right to work and to vote—and for greater control over their private lives and health—child-spacing and family planning—it reaffirmed women's place within a male-dominant domestic realm. This family form traveled the globe as part of the modernization discourse. The gender binary of the nuclear family in particular was said to provide the context for development of modern personality traits: individualism, acquisitiveness, achievement orientation, and a focus on the future. These purportedly American character traits and cultural competencies were said to be necessary to sustain modern progress individually and nationally; they were thus required of "backward cultures" of the Third World if they were to succeed as independent nations.[58]

Within this "context of discovery,"[59] demographic analyses of the first censuses conducted in the Global South using the United Nation's new standards announced a new population problem: "too rapid," indeed, "explosive" population "growth" in the "undeveloped world."[60] In a now quite familiar narrative, Western demographic analyses "found" that a dramatic decline in death rates was under way in emerging nations. That decline, coupled with the continuing high birth rates said to be characteristic of "traditional cultures," was causing rapid population growth. These analyses concluded that the current rates of population growth were excessive and projected dire effects of "too rapid growth" well into the future. The colonial difference—population growth before development—made prospects for the future even worse. Demographers provided precise numerical portraits of what was described as the devastating consequences of population growth on economic development. The numbers generated in demographic studies, and the urgency they conveyed, suggested that Malthus had been right when he concluded that unrestrained population growth threatened famine, war, and misery. The modern neo-Malthusian threat was that the market economies of emerging nations might succumb to famine, war, or Communist takeover.

The demographic problematization of population growth suggested a ready-made solution to the problem of excess fertility. Emerging nations—particularly in Asia, where demographers measured the greatest growth potential—needed to imitate the example of the West. Declining mortality should be followed by adoption of the small-family norm and a shift to low fertility, what demographers termed the *demographic transition.*[61] While the "original" Western demographic transition, like its economic development, was said to have occurred as a natural, nearly automatic, result of its modernization, the rest of the globe required expert guidance to achieve it. Demographic figures thus opened new possibilities for social action and warranted novel administrative attention and intervention into procreative processes. The logic of transition theory dictated that the best means of defusing the population bomb was effective fertility management guided by demographic expertise.

Transition theory's assumptions about high fertility in traditional cultures derived from contraceptive effectiveness studies of the 1930s. Those studies, which co-configured the binary of natural (uncontrolled) and controlled fertility, defined women as excessively fertile by nature. In contrast, controlled fertility was defined as a "cultural competency" of modern industrial life, at which white middle-class women excelled.[62] Representing the (recent) contraceptively produced small (white Western) family as the norm, demographic measures construed the population patterns of emerging nations in the Global South and the fertility of poor women of color everywhere as aberrant. Population figures simultaneously confirmed the superiority of Euro-American gender relations. Euro-American women who used contraceptives to "tame chance" and master their reproductive fate served as the standard-bearers of modern motherhood.[63] Euro-American men stood as the exemplars of modern masculinity because they "allowed" "their" women to exercise self-mastery. Demographic discourse positioned population control advocates in all geopolitical locales as securing a better future by rescuing women and cultures "mired" in nature.[64] In this way, concessions to the demands for reproductive control by Euro-American women quickly became new requirements of contraceptive control by poor women of color everywhere and population control by nations of the Global South.[65]

As demographers participated in national and international conferences and published technical reports detailing population facts, nation by nation, the demographic transition conceptualization and its numerical figures circulated widely among scientific, governmental, and popular audiences. The demographic account of population did not go unchallenged by masculine rivals, however. In successive UN conferences, representatives from socialist and non-aligned nations challenged the underlying economic logic of the population problematization; the Vatican and representatives of Catholic nations challenged its modernist vision of fertility and family planning. These debates represented struggles among competing reproductive masculinities over the meanings of modernity.[66] Yet, although family planning may have remained controversial, all sides accepted the facticity of the numbers and expressed anxiety about rapid population growth. Moreover, the demographic transition was institutionalized within the grand narratives of social scientific analysis and foreign aid projects.[67] Demography's figures of fertility and population growth thus became a taken-for-granted part of the mid-twentieth-century social imaginary.

Demographic Social Worlds

Demographers will likely object to the characterization of midcentury demographic facts as contributing to the panic about population. Disciplinary histories written by and for demographers treat population control programs as political matters that were and are independent of demographic knowledge products.[68] In particular, demographers are likely to object to my use of Ehrlich and Moore to introduce demographic reasoning. It is not so much that their numbers were wrong, but rather that the narrative attached to them used intemperate prose. Such works would be judged by disciplinary standards to have more in common with shrill political propaganda than with the discipline's own highly mathematical monotone.[69] In fact, disciplinary histories stress that the field has been repeatedly compelled to insulate itself from those who would appropriate its knowledge products for political purposes. Indeed, they argue that mathematical rigor became the primary means by which the discipline distinguished itself from population politics; some lament that it was nonetheless captured by the population control establishment.[70] Even recent critical histories of the field that acknowledge the "linguistic work" that demographers "accomplish[ed]" through their "fixing" of categorical standards still treat the numbers as separate from the politics that shaped the "market demand" for demographic knowledge products.[71]

The present analysis draws the line between demographic science and population politics differently. As Thomas Gieryn has noted, "Epistemic authority exists only to the extent that it is claimed by some (typically in the name of science) but denied to others."[72] The distinction demographers made between demographic science and population politics is part of the boundary work by which they created the social space for and credibility of their discipline, and by which it denied credibility to others. The assertion that mathematics enabled the field to insulate itself from the politics and enthusiasms of its time is itself a political claim. As the historiography of statistics has shown, relying on mathematics as proof of "the facts" exercises a potent modality of power, and, I contend, a deeply gendered one. Mathematics, in particular, represents the pinnacle of abstract rationality attributed to hegemonic masculinity.[73] That is, because demographic calculations are mathematically accurate, the numbers constitute powerful evidence of the field's epistemic rigor and dispassionate objectivity. Demography's various audiences may debate the completeness of the numbers, but all sides share a realist perspective on them. That is, they share the belief that the censuses and vital statistics that constitute population facts are based on simple mechanical acts of counting people and events that are "out there" waiting to be counted. Thus, there is shared belief that there is a real number to be found, and the identity and positionality of the counter is irrelevant to ascertaining it.[74] When the math works, when figures are taken for granted as reflections of the ways things are in the world, the modalities of investigation, the network of epistemic allies, and the inferential techniques by which the local politics and anxieties are inscribed in/by the numbers disappear from view. The complex and contingent social processes[75] and ideological frameworks upon which the reality of the numeric portraits depends are invisible, obscured by the "trust in numbers" characteristic of the statistical age.[76] Instantiated by statistical reasoning, demographic figures operate as independent material facts that demographers can report to scientific peers, policymakers, and students in "the flat voice of positivist social science."[77] This voice, I argue, is (not incidentally) a white Western man's voice.

Demographic boundary work was central to its wider cultural reception and also shaped the wider reception its rivals received. As Thomas Gieryn observes, "As a cultural space science cannot be under[stood] apart from the ground against which . . . it now appears in relief."[78] Thus boundaries define not only the inside but also the outside, the "domain of abject beings," those whose voices, perspectives, and claims can be discounted, derided, and ignored.[79] The following chapters demonstrate that in the case of demography, the women-led birth control movement served as the primary other against which hegemonic masculine demography defined itself but could not do without.

Like so many scientific arenas, demography was a strongly gender-differentiated space. According to one of the only female demographers of the period, "it was truly a gentleman's club."[80] This is not to say that there were no women who worked within the demographic organizations. There were; they generally held support roles—secretaries, technicians, and frequently wives—doing the clerical work involved in producing demographic charts and graphs. Yet, because demographers singled out women's advocacy on reproductive matters as exemplary of the political menace the discipline faced, gender became an explicit marker dividing demographic science from population politics. With it, demographers leveraged gendered signs of reason and emotion to insulate their practices from political scrutiny. Demographers and their professional organizations utilized the mantle of mathematical rigor to keep their distance from the birth control movement even as they collaborated in the social networks that configured the population control establishment. The US and international birth control movements stood alternatively as demography's abject others, rivals, respectful audiences, collaborators, and sometimes even as employers, but always as inexpert and sentimental outsiders. By fastening the boundary between their science and population politics to masculinity and mathematics, demographers set up a "troubled relationship" with feminism at its foundation.[81]

At the same time, demographic social worlds exercised the coloniality of power through complex relationships with scholars and family planning officials in the Global South. For instance, graduate student fellows from the Global South, mostly men, served in support roles in US demographic research agencies.[82] Student fellows often became such officials after completing their studies. Indigenous statisticians, US trained or otherwise, also participated in demographic social worlds as local experts who collaborated with and/or challenged US demographers. Their terms of inclusion within global demographic circles, however, positioned them as less authoritative.[83] Overall, subaltern demographers spoke in a (masculine) language that expressed confidence in the general theories and statistical techniques of the discipline. However, the basic Malthusian equation of poverty and population was a contentious matter internationally, and American demographers recurrently articulated the challenges of negotiating the "sensitive problems of public relations" involved when instructing subaltern citizens who might object to outsiders prescribing solutions to local problems.[84]

My analysis thus centers on US demography because it dominated the social-epistemic circuits that organized global flows of midcentury population crisis discourse and practice. But the genealogy proceeds by juxtaposing mainstream US demography with two alternative, rival discourses of reproductivity: the discourse of women's contraceptive need, advanced by the metropolitan woman-led birth control movement, as represented by groups associated with Margaret Sanger and the British birth control movement; and the anticolonial population discourse produced by Indian nationalists in the late colonial period. These rival epistemologies are only two among a number of possible comparators. I have chosen them because they are central to the local politics and worldly practices that shaped the trajectory of population knowledge and control. In the United States, the birth control movement was the strongest rival for funds, credibility, and influence faced by the emerging field of population science. India is only one of multiple colonies and histories implicated by the global reach of demographic figurations. But on the world stage, it is an important case because of the central place India holds in European colonialism, its demise, and the subsequent postcolonial development regime. In addition, in the West, as Ehrlich's vignette suggests, India exemplified overpopulation, even though it was the first nation to initiate population control through family planning.[85] It is therefore a valuable site for investigating the demographic crisis figurations that moved a "developing" nation to act. Comparison of these alternative narratives at the borders of dominant demographic discourse brings into sharper focus the specific contours of the gendered geopolitics inscribed within demographic theory and measurement practices.[86]

In particular, the analysis traces the alignment between demography and racial politics of the midcentury. The distancing of science and statistical expertise from population politics, as well as from the women-led birth control movement, conceals demography's active engagement with eugenics. That is, eugenics enjoyed a far different relationship to demography than did birth control, one primarily characterized not by distance, but by embrace. Even as birth control advocates were excluded, prominent eugenicists were welcomed into demographic circles, and demographers participated in eugenic organizations. They shared a commitment to statistical reasoning. They also shared the opinion that, on one hand, the birth control movement posed a menace to population science and, on the other hand, unrestrained population growth jeopardized society. The alliance of demography and eugenics grew out of efforts to constrain the effects of both threats. It was a conditional embrace, however. Demographers actively disputed genetic explanations of population trends that eugenicists popularized in the 1920s and 1930s, favoring cultural explanations instead.[87] The present analysis shows that the history of demographic epistemology was fundamentally shaped by the efforts to move population science beyond biological determinism and yet to secure ground for eugenics principles of better breeding.[88] Following from Balibar's discussion of cultural differentialism, cited above, I argue that the naturalized social history of population inscribed in demographic transition theory incorporated a version of eugenics in its conceptualization of cultural difference. The population crisis problematization, I contend, consituted a means by which the eugenic mission could carry on after eugenics itself was discredited.

Feminist demographers have applied a feminist lens to demography to critique the effects of its overreliance on positivist methods. Scholars such as Susan Cott Watkins, Susan Greenhalgh, Harriet Presser, and Nancy Riley have shown that demography, in its commitment to quantitative methods, has both included and ignored gender.[89] That is, demographic discourse includes gender as a nominal "attribute of individuals" and as a biological categorization of reproduction roles. In so doing, classic demographic theory allows no room for gender as a "structuring principle of social life," a set of social processes and power relations by which bodies are sorted and authority, rights, and resources are distributed. Thus, even though women's bodily processes constitute the primary object of fertility knowledge, women's multiple and diverse perspectives and agency are absent from demographic narratives.[90] These feminist scholars conclude that the field's dependence on positivist mathematics and its nominal gender concept result in a troubled relationship with feminism. Feminist demographers also recognize that complex relations of gender, ethnicity, and geopolitics intersect in the field's history.[91] They tend to treat these factors, however, as occurring independently and then intersecting, in sequence. As insiders, feminist demographers have been most concerned with providing effective correctives to the twin problems of the field's flawed approaches to gender and their own marginalization. Written for demographers, feminist critiques of their discipline's history are offered in service to their larger and more important claim, which is that demographic research must "address the full spectrum of social, political, and economic forces that influence demographic behavior."[92] Their accounts tend to concur with disciplinary histories in concluding that demography's credibility required that it insulate itself from politics.[93] Moreover, while they effectively delineate the empirical limitations of demographic theory, they nonetheless maintain a strong investment in their discipline's "realist" perspective on demographic numbers.

This analysis builds on the very valuable work of feminist demographers, but takes an agnostic perspective on the demographic facts.[94] That is, drawing on recent historiography on statistical regimes of knowledge within the European metropole and colonial contexts, I understand demographic facts to be the products of population theory, a conceptualization of biosocial reality, not a reflection of it.[95] To the extent that they represent the actual, they do so configured in the naturecultural terms of that theory.[96] Furthermore, because population figures are the products of theory—a representation, not a reflection, of reality—they can be analyzed just like any other rhetoric to illuminate the grammar that organizes them, the string of associated signs that give them meaning, and the connotations and affects that adhere to them.[97] The historiography of statistics focuses on the social and political processes of quantification, particularly censuses. It demonstrates that quantification involves an exercise of power that denies it is any such thing. It also illuminates the various ways in which statistical regimes of knowledge have permeated the modern social imaginary, shaping our apprehension of the social and natural world. However, this historiography has tended to overlook gender politics. Expanding consideration of the politics of numbers, I contend that gender is a constitutive element of social statistics, and that vital statistics in particular are an important site where gender binaries and hierarchies are assembled and their politics become entangled. Moreover, I am particularly interested in the gendered grammar that organizes objects and subjects of population discourse. It may be obvious that gender differences and hierarchies configure objects of attention in the classificatory systems constructed in midcentury vital statistics.[98] But, as this analysis demonstrates, the gender of the knowing subject of statistics, the mathematically skilled masculine subject, the Malthusian man, was also configured in statistical regimes of the midcentury population crisis as the authoritative expert on dynamics of human reproduction.

My aim is not to dismiss demographic figures as simply wrong or as mere fabrications, but instead to excavate their history in order to highlight the "host of inferential technologies" by which the truth of midcentury population statistics was produced and their credibility maintained.[99] Because, as Theodore Porter has observed, "numbers work most effectively in a world they have collaborated in creating," it is crucial to investigate what cultural and affective work the figures of the population explosion accomplished.[100] The tools of textual and cultural analysis offer a rich method of investigation. They focus attention on the subjects, objects, events, and affects with which demographic figures populate the world. They open space to ask if those figures are reasonable approximations of the actual. And to ask how the power relations that underlie their making lead to distortions.[101] In answering these questions, I contend that entangled politics of gender and racialized colonial difference impelled the numbers' cultural work of installing effective contraceptive practice by women as a mandatory cultural competency and making low fertility the "vital" marker of the modern citizen and nation.[102]

The following chapters elucidate the interplay of gender binaries and hierarchies in the production, circulation, and repetition of mid-twentieth-century demographic facts and figures. They focus on how the literary codes, conventions, and regulatory hierarchies of disciplinary texts configured "whole strings of associations and carefully nuanced stories" about the components of the overpopulated modern world and invited "tutored" audiences to take up positions within that world.[103] The demographic facts and figures circulated through an affective economy of gendered and geopolitically specific anxiety for and about the future; repetition of those facts and figures amplified that anxiety into a math panic, stoked fear of a cataclysm, and moved individuals and nations to act.

The gendered coloniality of the midcentury population imaginary continues to haunt the terrain on which feminist activists in the Global North and South have endeavored to work in coalition. Holding Western family planners accountable for their participation in the population control establishment has played a vital part in working through the distortions and injustices within this terrain. By excavating the history of the demographic fact-making that stoked the population panic and directed population control, the figures of which continue to populate the world, I seek to deepen feminist understanding of the dimensions of reproductive injustices. In this way, I seek to move feminist understandings of reproductivity away from attachment to figures that support continued state-sanctioned interventions into women's reproductive lives. Reproductive justice advocacy will always need to engage with aggregate statistics both to contest injustices and to assess efforts to redress them. Thus, it is important to understand how the numbers have worked to frustrate those goals.

Structure of the Book

Chapter 2 begins with the famous *Essay on the Principle of Population,* by Thomas Malthus, whom mid-twentieth-century demographers identify as the founding father of the discipline. Reading the *Essay* as an account of the reproductive mastery of bourgeois European masculinity, the chapter illuminates the intertwined logics of gender difference and hierarchy that organize modern population discourse. The chapter then provides a cultural-epistemic history of twentieth-century demography's principal object of attention—population dynamics—in order to delineate the agnostic perspective on statistical styles of reasoning used here.

Chapter 3 focuses on the configuration of gender difference and hierarchy by which demography defined the boundaries of legitimate population knowledge. Excavating the "official story" of the discipline's founding event, the establishment of the Population Association of America (PAA) in 1931, the chapter shows how demographers deployed gendered tropes to capture authority for themselves and deny it to others. The chapter contrasts the exclusion of birth control advocates from demographic circles with the embrace of eugenics by demographers. Using the example of Frederick Osborn, a leading eugenicist and the first executive director of the Population Council, the analysis traces the affective and epistemic alignments between demographic and eugenic conceptualizations of population problems.

Chapter 4 focuses on the gendered binary that organizes the demographic conceptualization of fertility. Through close readings of early contraceptive effectiveness studies, it interrogates the intellectual apparatus with which demographic theory defined pregnancy risk and calibrated natural and controlled fertility. It shows that the figures of natural and controlled fertility enabled demographers to construct a hierarchical scale of femininities that arrayed contraceptive competency by purported race and class deficiencies. By juxtaposing demographic measures with women's narratives of pregnancy risk and contraceptive need, the chapter also elucidates the erasure of gendered tensions in demographic figures of reproductive risk.

Chapter 5 shifts the analytic focus to masculinity hierarchies configured in population knowledge through their alignment to feminine fertility figures. In particular it excavates the gendered coloniality of the demographic theory of population change—the demographic transition—focusing on the work of Frank Notestein, the founding director of the Princeton Office of Population Research (OPR). The chapter continues with a close reading of population policy debates conducted by the Population Council that delineates the gendered geopolitical contours of the council's approach to disseminating demographic knowledge and the population control objectives it authorized.

Chapter 6 continues the focus on hierarchies of masculinities through a case study of population knowledge in and about India. It displaces the Orientalist view of India as a beneficiary of Western expertise by examining the internal Indian debates about population in the later years of colonialism and early years of independence. Through a close reading of population analyses produced by Indian demographers, the chapter delineates the intellectual and affective alignment of population discourse with Indian nationalist politics. The chapter juxtaposes this account with midcentury demographic representations of India as the exemplar of the population explosion in demographic texts that set the terms of midcentury debate about the population crisis.

The concluding chapter turns to the question of the relationship between demography and transnational feminist reproductive politics by assessing the fate of demographic knowledge and goals in the 1974 and 1994 UN population conferences. In both cases, the priorities of population control were displaced by counternarratives of anticolonial development and women's rights. However, the demographic figuration of population change remained authoritative. The chapter concludes with a call for a new regime for building trust in numbers, one that uses a transnational feminist science studies lens to take full account of the historicity of their calculation and the cultural and affective work they perform.

2

Rereading Malthus

Population and Masculine Modernity

> Population is not an observable object, but a way of organizing social observations. The concept emerges out of projects that seek to configure social relations so that these may be known and mastered.
>
> —BRUCE CURTIS

THOMAS MALTHUS, WHOM MID-TWENTIETH-CENTURY DEMOGRAphers proudly claimed as their founding father, published the first edition of *An Essay on the Principle of Population* in 1798. Offered as a refutation of William Godwin and Jean Condorcet's vision of the improvement of society through progress, Malthus predicted a less sanguine human future.[1] In five successive editions, Malthus advanced his argument by incorporating more numerical data and charts to sustain his claim that the general law of population would prevent permanent advancement in the human condition. Critics characterize Malthus as a dismal philosopher, pointing to dramatic improvements in agriculture and industry as proof that he underestimated the human capacity to expand nature's limits.[2] Regardless of the accuracy of its predictions, Malthus's *Essay* is important because, as Mary Poovey observes, it "constitute[s] a crux in the history of the modern fact." Through its several editions and the controversy surrounding them, Malthus helped "rework the cultural connotations of numerical representation" to "stress their impartiality and methodological rigor" as a way to transcend politics.[3] Moreover, Malthus's *Essay* circulated widely throughout the nineteenth and twentieth centuries, influencing such thinkers as John Stuart Mill, Charles Darwin, and John Maynard Keynes.[4]

In laying claim to Malthus, twentieth-century demographers staked their disciplinary origins to his quantitative reasoning, which provided a rich resource for establishing their own impartiality and for blunting political scrutiny of their claims. But Malthus offered more than mathematical positivism. He brought mathematics, hegemonic masculinity, and methodological rigor into association in the modern discourse of population. That is, the knowing subject of Malthus's *Essay*, the rational individual who deduced the principle of population, is an exemplary English gentleman who stands atop the scale of men and nations calibrated to vital statistics. Rereading Malthus as a narrative of hegemonic masculinity brings into focus the binary and hierarchical relations of gender organized in and through population reason. It also brings into sharper relief the patriarchal imperial landscape inscribed in his universalizing narrative of population, a narrative on which mid-twentieth-century demographers would expand.[5] Finally, Malthus's *Essay* exemplifies the privileged position of statistical regimes of knowledge in modern European nation-states and their imperial enterprises. The chapter therefore takes a closer look at the statistical reasoning inscribed in the demographic concept of population and the social facts assembled by it. In particular, the discussion elucidates the co-configuration of the individual and aggregate and the slippages between them that has made population such an effective tool of governmentality.[6]

Rereading Malthus

The *Essay* presents its case in a classical positivist format. The first chapter enunciates the general law of population. This philosophical discourse is followed by a survey of world population that amasses data about agricultural productivity, marriage, birth, and death. The final chapter restates his initial reasoning, concludes that the evidence confirms his proposition, and responds to Godwin, Condorcet, and his critics. The assembled evidence ranges from the purely narrative accounts written by colonial agents and missionaries to official statistics of European states. As the survey proceeds, the balance shifts from narrative to numeric evidence. Thus Malthus's march through the evidence is simultaneously a march toward modernity based in the impartiality and rigor of numbers.

Malthus's basic proposition is that any sustained improvement in the human condition is unlikely because different (natural) mathematical functions underlie population and resource growth. He reasoned that plants and animals "are all impelled by a powerful instinct to the increase of their species," and "wherever, therefore, there is liberty, the power of increase is exerted," leading to "superabundant effects." The underlying mathematical law is that population increases geometrically, as "the numbers, 1, 2, 4, 8, 16, 32, 64, 256" do. Because "man" is one of nature's creatures, he is subject to the same laws of increase. The European settler colonies of North America offered the best example. With unrestrained fertility levels and an ample food supply, the US population was said to have doubled in only twenty-five years.[7] However, unrestrained population growth puts enormous pressure on the resources needed to sustain it. Food resources, in particular, Malthus asserted, are added more slowly, "when acre has been added to acre," through "human industry." Moreover, growth of food resources ultimately is limited spatially. Even under "the most favorable" circumstances, Malthus asserted, "the means of subsistence" can increase only arithmetically "as 1, 2, 3, 4, 5, 6, 7, 8, 9" do.[8] Thus, "population has this constant tendency to increase beyond the means of subsistence." The mathematics represents the inherent conflict between the nature of man's passion and his industry. The drive to propagate, which is the result of passion and instinct, differs in kind from the impetus to increase material resources, which derives from "science" and "industry."[9] These different forces cannot easily be reconciled.

For Malthus, the balance of population and resources was maintained by the "positive checks" of disease, famine, war, and vice or by the "preventive check" of reasoned moral restraint.[10] Man's capacity for reason offered the one hope for transcending nature's ruthless means of stabilizing these countervailing forces. Man possesses "that distinct superiority in his reasoning faculties, which enables him to calculate distant consequences." Therefore, although "man" is "impelled to the increase of his species by an equally powerful instinct, reason interrupts his career, and asks him whether he may not bring beings into the world for whom he cannot provide the means of support." As such,

> he cannot contemplate his present possessions or earnings . . . without feeling a doubt, whether if he follow the bent of his inclinations, he may

> be able to support the offspring which he will probably bring into the world. . . . Will he not lower his rank in life, and be obliged to give up in great measure his former society? . . . Will he not at any rate subject himself to . . . more severe labor? . . . Will he not be unable to transmit to his children the same advantages of education and improvement he had himself possessed? . . . And may he not be reduced to the grating necessity of forfeiting his independence and of being obliged to the sparing hand of charity for support?[11]

Such calculations would lead reasonable men to restrain their passion. By this logic, sexual economy trumped political economy in determining the level of misery and want in any society. A man's status within the ranks of his fellows was a consequence of his ability to contain and appropriately direct his (heteronormative) desire to his best economic advantage.

As the above quotations make clear, the subject of the principle of population is masculine. This figure embodied a specific, exemplary, bourgeois masculinity that tamed passion with reasoned self-discipline. For Malthus, the chaste delay of marriage was the only morally acceptable restraint on that passion. The Malthusian man applied market principles to calculate the cost of children before embarking on his marital career. In so doing, he reversed nature's grip on him. The modern nations of Europe were the only ones in which the preventive check of prudential restraint of sexual expression had loosened nature's hold on society, at least among the middle classes. Malthus specifically described the instinct to propagate in (heteronormative) gendered terms when he noted that "whether the law of marriage be instituted or not, the dictate of nature and virtue seems to be an early attachment to one woman."[12] This description gives male-dominated (heteronormative) marriage the "alibi of nature": man's natural virtue demands (bourgeois, heteronormative) marriage. Moreover, Malthus also asserted, "The passion between the sexes has appeared in every age to be so nearly the same that it may always be considered, in algebraic language, as a given quantity."[13] No further analysis was required. By nature and the numbers, sexual desire could be reduced to a universal algebraic constant.

The failure to restrain passion within reasonable limits carries grave risks. Malthus warned that "promiscuous intercourse, unnatural passions, violations of the marriage bed and improper acts to conceal the consequences of irregular connexions, are preventive checks that clearly come under the head of vice," and "the general consequence of vice is misery." Illicit sexual expression degraded "in the most marked manner the dignity of human nature," but in gender-specific ways. "It cannot be without its effects on men, and nothing can be more obvious than its tendency to degrade the female character, and to destroy all its most amiable and distinguishing characteristics." In turn, the virtue of society depended on the character of men's relations with women, because libertine women necessarily are inadequate, careless mothers. "Among those unfortunate females . . . more real distress and aggravated misery are perhaps to be found than in any other department of human life." Malthus thus concluded, "No person can doubt the general tendency of an illicit intercourse between the sexes to injure the happiness of society."[14] Women entered the narrative as the object of masculine passions. They are "the sex" with whom men had either virtuous or vicious relations and their character was determined as a result.

Malthus's view of the gender binary reflects the conventions of his time. Through marriage, the modern European nation-state configured the fiction of the public and private worlds that sustained itself, the public order, the modern citizen, and his freedom in the male-headed family unit, the basic unit of society. In this formulation, sexuality, the married woman, her citizenship, duties, and freedom are subsumed in familial obligation.[15] With this gendered fiction, contested gender relations between women and men were bracketed as the personal troubles of a man and his wife. William Godwin, one of Malthus's interlocutors, expressed the effect of this formulation with regard to procreation: "It is one of the clearest duties of a citizen to give birth to his like, and bring offspring to the state. Without this he is hardly a citizen."[16] The masculine pronoun in connection with giving birth is stark evidence of the masculinism of the configuration on gender, generation, and nation inscribed in population reason. For both Godwin and Malthus, the problem of procreation as a governmental concern emerges only after the operations of gendered power that coupled women to men. It is a problem of managing relations among men, of whether men are measuring up to the procreative standard, and, thereby, their duty to the nation.[17]

While there has been a lot of attention focused on Malthus's use of revised editions of the *Essay* to oppose British poor laws, far less attention has been given to his participation in the European imperial project. From 1805 until his death in 1834, Malthus lectured on political economy at the East India Company's College at Haileybury, where he helped prepare young men to serve as the company's agents in India.[18] From this vantage point, Malthus's participation in and knowledge of emerging practices of empire paralleled his contributions to them.

The first book of the *Essay* consists of a wide-ranging survey of the "immediate checks to population in the past and present state of society."[19] That survey provides the background horizon for his discussion of modern Europe in book 2. A common gesture of colonial difference, temporal distancing, structures the order of discussion in book 1. That is, the chapters map time onto geographic space. As it journeys around the globe, the survey simultaneously journeys back in time. Distance, as a spatial measure, thus also marks the temporal progress from primitive to modern man.[20] For example, book 1 is titled "Of the Checks to Population in the Less Civilised Parts of the World and in Past Times." Chapter 1 discusses groups at the "lowest stages of human society" (the inhabitants of the Americas and the South Sea Islands). Subsequent chapters discuss the ancient societies of Europe (those Germanic hordes who toppled Rome) and the current societies of East and South Asia (India and China figure prominently). His survey ends with consideration of the classical worlds of Greece and Rome, which, according to the conventions of European epistemology, Malthus positions as the immediate antecedent of modern Europe.[21]

The juxtaposition of Europe's ancient past with the current societies of the rest of the world constructed a single universal scale with which to array humanity. As Walter Mignolo has noted, "Arrangements of events and people in a time line is also a hierarchical order." "As a principle of order," time "relegat[es] them to before or below from the perspective of [its] 'holders.'"[22] In this case, by likening them to Europe's ancient past, Malthusian time positioned current Asian societies as archaic. They are both timeless—their past, present, and future are the same—and in contrast to modern Europe, they are always already behind, belated, and benighted. The absence of numeric representations amplified the past and the primitive position as "behind the times."[23] It was as if their social relations were not advanced enough to be enumerated. Conversely, European superiority was emphasized through the dense numeric representations available to describe them. England, where preventive checks were said to operate "with considerable force," had so much data it required two chapters. The final nation surveyed, he described it as one in which no "extraordinary mortality" or wide variations in the proportions of birth and death were seen, indicating that "habits of prudence . . . prevail."[24]

Where numerical data were lacking, travel literature and colonial reports sufficed.[25] For instance, Malthus relied on Captain Cook's memoirs for his description of "the wretched inhabitants of Tierra del Fuego," whom he placed at the "bottom of the scale of human beings." He said, "Of their domestic habits and manners, however, we have few accounts. Their barren country, and the miserable state in which they live, have prevented an intercourse with them that might give such information; but we cannot be at a loss to conceive the checks of population among a race of savages, whose very appearance indicates them to be half starved, and who, shivering with the cold, and covered with filth and vermin, live in one of the most inhospitable climates in the world."[26]

Gender relations are central to these accounts. Malthus relied on a catalog of sexual customs described in such sources to estimate the level of civilization attained by the men and nations in his survey. His descriptions participated in the standard fare of colonial discourses on sexuality. For instance, in talking about the "races of savages," he highlights violence as a prelude to sex in rather lurid descriptions of abduction and abuse: "He steals upon her in the absence of her protectors, and having first stupefied her with blows of a club, or wooden sword, on the head, back, and shoulders, every one of which is followed by a stream of blood, he drags her through the woods by one arm, regardless of the stones and broken pieces of trees that may lie in his route, and anxious only to convey his prize in safety to his own party, where a most brutal scene ensues. The woman thus ravished becomes his wife."[27] "Thus ravished," women's finer qualities are destroyed, which leads them to a life of promiscuity and bad mothering.[28] In describing India, Malthus noted that the restraint on growth one might expect to result from premarital chastity was undercut by India's strong cultural inducements to early marriage. His narrative of Indian manners and customs included a list of affronts to women, which resulted in their "mutable temper . . . want of settled affection, and perverse nature." He concluded, "India, as might be expected, has in all ages been subject to the most dreadful famines."[29] Malthus attributed the vast population of China to "the persevering industry of the Chinese" and "extraordinary encouragements that have been given to marriage." He argued, "The marriage union, as such, takes place in China wherever there is the least prospect of subsistence for a future family." Malthus gave credit to virtuous gender relations, noting that "the women are said to be modest and reserved." However, "notwithstanding the great sobriety and industry of the inhabitants of China," the overall picture is one of severe labor caused by the lack of reasonable restraints on marriage. Noting the history of famine in China, he predicted that famines would "be more severe" in the future.[30]

Thus, book 1 instantiates the bourgeois colonial discourse of self-mastery, which, as Ann Laura Stoler notes, established "certain cultural competencies, sexual proclivities, psychological dispositions, and cultivated habits [that] defined the hidden fault lines—both fixed and fluid—along which gendered assessments of class and racial membership were drawn."[31] For Malthus this discourse of self-mastery calibrates masculinity and civilization by imperialist measures of sexual civility and crude death rates. In the aggregate, he reasoned, poverty, misery, and want resulted from the failure of men to maintain virtuous sexual habits and customs; thus mortality rates offered a particularly reliable marker of the level of moral restraint exercised in a population. Where morality was high, mortality would be low; where morality was low, mortality would be high.[32] The Malthusian narrative, coupled to numerical figures, assembled the bourgeois European balance of productive and procreative forces as the ideal. It was a modernity of sustained progress and industry that rested upon the reasoned judgment of the bourgeois masculine subject whose wise investment in male-dominant (heteronormative) marriage organized the appropriate relations between the sexes. Deviations from the bourgeois standard marked each group's distance from the normative ideal of civilization. Women were positioned as tokens of men's individual and aggregate virtue. The Malthusian scale thus lent quantifiable support to the imperialist conceit that the level of civilization could be judged by how each society treats "its" women. At the same time, the focus on the sexual economy as a timeless feature of the indigenous landscape diverts attention from the effects of colonialism. By Malthus's calculations, famines and epidemics were nature's checks on sexual imprudence—not the result of colonial economic exploitation and political subjugation.[33]

Population Statistics and the Nation

Malthus concedes that actual observation of the action of the population principle was limited, noting that the "oscillation" of "retrograde and progressive movements" was "less decidedly confirmed by experience than might naturally be expected." This was because "the histories of mankind which we possess are, in general, histories only of the higher classes," and it was among "the poor" that the "constant check upon population acts."[34] Thus, those most affected by the general law of population were least likely to record it. To overcome this observational difficulty, Malthus encouraged the development of statistical techniques to measure key operations of the population principle, such as marriage rates, birth rates, and death rates, at the societal level.

With his call for the use of statistics, Malthus's *Essay* aligned with instrumentalities of modern governmentality. Focused on the "administration of bodies and the calculated management of life," European (masculine imperial) nation-states developed statistical styles of reasoning as one tool "for achieving the subjugation of bodies and the control of populations."[35] Population statistics became a way to focus and extend an "administrative gaze"[36] as those nation-states endeavored to govern expanding territories filled with dynamic varieties of "bodies capable of performances."[37] Quantification practices made sense of the chaotic diversity of bodies, events, and goods by aggregating them into numeric representations in tables, charts, and graphs. From the "avalanche of numbers" that characterized European metropoles and colonies, "governments perceived" that they were dealing with a reality previously obscured by the details of the particular.[38] That new object of knowledge, "population," was, according to Foucault, a unique entity with "specific phenomena" and "peculiar variables" of its own, all of which were "situated" at the intersections of "characteristic movements of life" and "specific effects of institutions."[39]

The statistical style of reasoning "has come to sustain a remarkably powerful way of gaining purchase on dimensions of social life."[40] The figures it creates permeate the modern social imaginary, shaping our apprehension of the natural and social worlds. Counted and cross-tabulated, the characteristic movements of life are decontextualized from the details of particular incidents and recontextualized as numerical relations that "display the qualities of generality and permanence."[41] In this sense, statistical reasoning about population transformed chance events into social regularities, knowable through the "cool idiom of numbers." So known, they could be acted upon. Statistics thus generated a "sense of a controllable indigenous reality" and configured new objects of knowledge, "matters of national importance,"[42] opening vast new spaces for technologies of power to regulate social phenomena.[43] The transformation of the characteristic movements of life into statistical knowledge of population proceeds on two axes, one totalizing, one individualizing.[44] To illustrate these conjoined knowledge-building processes, consider the national census, an innovative statistical form of the late eighteenth century that extended the administrative gaze across vast expanses of nation and empire. The census classified, counted, and aggregated "pertinent"[45] bodies within a territory, producing a variety of new totalities: the social body,[46] the nation, the people, the general public,[47] mankind, and society itself. But it is also a means by which different totalities—genders, races, and ethnicities—can be quantified. Through censuses, social bodies are mapped within geopolitical territories to become part of the territory itself. Moreover, even as aggregation configures numeric portraits of the totality and various subgroups, comparisons of the features of particular individuals within them configure numerical representations of the average citizen and his various others.

As Bruce Curtis has demonstrated in his groundbreaking analysis of the first Canadian censuses, however, the social totality, population, is "a theoretical entity, not an empirical entity." Population is "not an observable object"; it is a concept that provides "a way of organizing social observations."[48] It is not a reflection, but a theorization of social reality, of bodies in time and space. That is, the central fiction of the census is that "everyone has one—and only one—extremely clear place" in which to be properly counted.[49] Census-making is represented as a simple matter of matching the individual's place on the map to the right place on the census form. However, as Curtis has shown, every census depends on a particular imaginary landscape of "human bodies" in "virtual time" and "virtual space."[50] In order to make those counts, censuses assemble in advance the "authoritative categorization"[51] of the pertinent elements—sex, age, race, caste, household formation, income level, marital and employment status, community type (urban, rural)—in terms of a historically specific set of assumptions about "who" is "out there" waiting to be counted. Inscription in the census form, therefore, is an interpretive act. It is defined by "observational protocols,"[52] specified in rules of equivalence by which to identify and differentiate like bodies and to quantify them appropriately in administrative time and space.[53] With these sets of equivalences, statistical representation "yoke[s] together" things that are otherwise separate.[54] In this way, circuits of power organizing authoritative categorizations and statistical reasoning "capture and appropriate" diverse lands and peoples within an imagined vision of social life that links bodies, signs, and figures into affective and epistemic economies.[55]

To the extent that observations of social relations hinge on identifying bodies in time and space, census-making is a powerful instrumentality. The census concept of proper places is inherently political. It involves both the normalizing judgment (you are in the right place) and disciplinary practice (you must take your proper place) that are the hallmarks of Foucauldian power.[56] As Arjun Appadurai has noted, the "manifest rhetoric" of census figures is "technical (that is positivist, transparent, and neutral), but [their] subtext is contestatory (in regard to superiors) and disciplinary (in regard to inferiors)."[57] The social processes that produce census figures involve local functionaries in pedagogical interaction with "residents" as they observe, interpret, and inscribe information in the already defined categories. As residents participate in the census, their local knowledges, identities, and categories are reconfigured and translated into categories intelligible to the administrative processes of the state.[58] The enumerated learn to occupy places and tell stories that make sense within the normative categories. They learn how to cast themselves into figures that populate the nation/empire. Stories that do not make sense often become noise in the data.[59] Consider age, a seemingly straightforward identity category. Its classification depends on formal timekeeping systems and birth registries to mark events in administrative categories. Vital statistics systems thus shore up census-making by providing official time and place stamps on birth and death. However, in their absence, it is difficult to situate age precisely in the timescale of population statistics. Extremely pertinent to the population projections in the 1950s and 1960s, improving the accuracy of age reporting in censuses was a major focus of UN-led standardization.[60]

The effects of authoritative categorizations are conditioned by the extent to which they are inscribed in administrative processes that shape the "life chances" of the categorized. Their effects are conditioned by the extent to which those so categorized take it up (or are compelled to take it up) as an identity, a way of living, a pattern of conduct.[61] As an example, what we think of as gender is thinkable in part because of the classification system of vital statistics, in which birth certificates designate membership in one of the two categories of the gender binary. That assignment endures throughout one's life and death and influences one's "life chances" along the way. Similarly, what we commonly think of as race and caste are thinkable in part because of the apparatus of population. For instance, India's caste categories were largely negotiated products of the classificatory procedures of British colonial officials that were refined and expanded in sequential censuses. Negotiations of the official identities involved many parties over time as groups vied for positions from which to define themselves and others within the colonial matrix of rank, rights, and obligations. Racial categorization in the United States has also been subject to decennial revision throughout its history.[62] Such negotiations take place within circuits of power that position some with greater resources, more authority, and stronger voices than others. Once settled into their numerical forms, the authoritative categorizations appear as natural objects. The supporting institutional infrastructure and inferential modalities that produced them are not visible.[63] In this sense, then, census inscription "makes up people," providing them with a genealogy, a future, and an essence.[64] These categorizations are not simply instrumentalities of rule, however. Social movements take up and bend social categorizations to their own purposes as well.

One of the first individualized figures produced in population statistics, the average citizen made his appearance in Belgium in the calculations of Lambert Adolphe Quetelet, who in 1825 used rudimentary census data to construct a portrait of the height and girth of the average Belgian (man).[65] Quetelet's calculations used the numeric average—a statistical artifact—to express typicality. The equation of the aggregate mean with the average individual has since become ubiquitous. Arithmetic figures of who "we" are (what we do, think, eat, possess) routinely parade through all sorts of public discourses.[66] Such numeric representations meld the aggregate mean, the typical, and the ideal, solidifying the descriptive average as the standard, the prescriptive norm. They exemplify the kinds of people who should and do belong. Quetelet's average man symbolized the essential qualities of the Belgian national character, combining cultural symbols and normative conceptualizations of manhood exemplified by "habits of propriety" and "temperance."[67]

The slippage between the aggregate mean, the average man, and the ideal citizen rests on the exercise of normalizing judgments, in the statistical sense. That is, Quetelet derived his figures by applying the mathematics of the normal distribution under the law of error to describe variations in human qualities. The law of error was devised by astronomers to account for errors in their celestial observations. Quetelet made novel use of this mathematics to assemble his observations of social difference. That is, techniques of statistical normalization represent the modal point in a scattered array of observations as the arithmetic mean. This numeric representation can be used, as Quetelet did, to describe that modal point of observations qualitatively, as a measurement of propriety.[68] In contrast, descriptions of the "other" observations can be represented in terms of their distance, or deviation, from the arithmetic average, the normalized figure of propriety, which Quetelet also did. "The implication of Quetelet's idealizations of the mean was that all deviation from the mean should be regarded as flawed, a product of error."[69] It should be noted, however, that while the errors in celestial observations resided in the observations, not the stars, in the case of human variations, the flaws resided in the observed. In contrast to the idealized average man, "the others," who deviated from the mean value of a pertinent characteristic, became deviant. Investment in statistical forms thus made it possible to configure numeric representations of both the ideal and abject based on a variety of identifying features.[70]

Selective disaggregation of the total population into authoritative subgroupings extended and amplified the power of statistical normalization. After censuses categorized and counted authoritative "others," statistical analysis could characterize their typical qualities and contrast them to the (idealized and prescriptive) norm. In the early twentieth century, eugenicists developed the full mathematical potential of statistical normalization for representing and analyzing social differences by class, race, and coloniality. In fact, the founders of eugenics, Francis Galton and Karl Pearson, contributed substantially to the statistical science Malthus had called for. In the late nineteenth century, Galton, who gave eugenics its name and underlying logic, developed the basic statistical tools of correlation and regression to address qualitative questions of heredity. In his twentieth-century biometric studies of heredity, Karl Pearson developed more fully the mathematical potential of his general theory of probable errors, "the normal curve," upon which much of modern statistics is based.[71] Extracting from the Darwinian assumption that lower animals had to be more prolific to survive misfortune and predation, Social Darwinist Herbert Spencer popularized the view that intelligence and prolificacy must be inversely related.[72] In 1912, Karl Pearson analyzed British census data and concluded that one-half of each generation was the product of one-quarter of the married population, and that quarter, he argued, consisted of a disproportionate number with low social and economic standing.[73] This difference in fertility patterns among different "classes" of people, called differential fertility, had long been a concern among "old-stock" Americans and the European bourgeoisies. In the twenties, fear of natural differences in fertility spurred interest in population science as well as immigration restrictions in Western nations.[74] The first generation of US demographers, most of whom were "old-stock" Americans, helped shape this national discourse about differential fertility rates and their potential to "degrade" the "American" population, society, and culture. By the thirties, with legal restrictions on immigration and the apparent decline of death rates achieved through public health reforms, the weight of individual women's and men's reproductive actions on aggregate population quality and quantity became the focal point of concern.[75] These concerns increased in the face of a rapidly declining birth rate during the Great Depression.[76]

Normalizing judgments that rely on comparing averages homogenize "us" in contrast to "them" by means that sharpen and amplify differences. For instance, sex-difference research generally compares only arithmetic means, the average differences between women and men. This practice highlights the differences between the genders while masking differences within each group. In this way, statistical reasoning both "recognizes and disavows" difference in hierarchical binaries.[77] Hegemonic administrative frameworks generally label the differences of the deviant as self-disqualification from the positions and privileges of the normative.[78] As an example, the mid-twentieth-century woman who bore either fewer or more children than the average was characterized as selfish and unwomanly or submissive and hypersexual, respectively.[79]

Normalizing judgments also configure numeric portraits of the population as an aggregate entity, a social body. Just as individual identities are the product of authoritative categorization, however, so are aggregate population characteristics artifacts of the administrative conditions and statistical processes that produce them.[80] Consider vital statistics, a statistical form invented in nineteenth-century Britain that cast life and death into rates.[81] By compiling lists of disparate instances of death by cause, location, and time, statistics produced a new object of knowledge, mortality rates. The application of statistical tools to arrays of observations "revealed" regular patterns in those rates. Thus statistical regularity normalized death, making the chances of dying appear predictable and providing a measure of the relative state of public health and national well-being.[82] However, "nature has no such rates."[83] Births and deaths occur, certainly, but statistical procedures give them uniform meaning and sort them by type. Statistics situate specific instances in relation to the totality as a ratio (number of deaths to total population). In addition, as with the normalizing judgments regarding individual bodies, vital rates of social totalities can be construed in distributions, producing melded portraits of the typical, the normal, and the ideal, in contrast to the deviant and the abject. Here again, rules of equivalency classify, sort, and rank disparate social bodies across time and space. Malthus's comparative mortality-equals-morality scale is an example of a normalizing index of social bodies according to the self-disqualifying grammar of colonial difference.

Statistical knowledge about the social body opened new avenues for governments to act on behalf of the totality to influence the prevailing conditions said to generate patterned rates. According to Foucault, in the "society of normalization," governmentality joins disciplinary instrumentalities directed at individuals to regulatory instrumentalities directed at the total population.[84] Public health campaigns, which grew along with statistical knowledge of morbidity and mortality, exemplify these modern instrumentalities. Disease control campaigns, such as malaria control efforts after World War II, joined new regulatory technologies targeting the totality of population (DDT to eradicate mosquitoes) with techniques for governing of the self (sleeping under mosquito netting). In and through campaigns that join disciplinary and regulatory instrumentalities, statistical normalization shapes the reality to which it refers.[85] In a process Ian Hacking refers to as dynamic nominalism, the articulation of a norm can shape behavior as people strive to meet the standard.[86] Such striving affects future measurements, as reported practices shift closer to the stated norm. Certainly knowledge derived from aggregation practices can be of enormous individual and social value when it facilitates coordinated responses to illnesses like malaria. Yet at the same time, statistics that configure portraits of deviance can provoke and sustain moral panics, which authorize administrative interventions into wider social processes to move population patterns toward the prescriptive norms.[87] The mid-twentieth-century population explosion narrative and resulting population control interventions incited a mathematically based moral panic.

The focus of attention within population statistics is historically conditioned.[88] Early social statisticians such as Quetelet and Malthus worked in the shadow of the French Revolution and Napoleonic Wars. They shared a commitment to use statistics to temper politics by guiding the cultivation of moderate habits among individuals, "a narrowing of the limits within which the 'social body' oscillated," and thus quelling "the risk of 'falling into an extreme.'"[89] Over the course of the nineteenth century, innovations in population statistics focused on mortality measures at a period during which mortality control was especially pertinent.[90] Disease surveillance, to which mortality statistics were central, conceptualized regularity as the normative state for disease rates. Change, especially change perceived as sudden and unexpected spikes, thus became non-normative, provoking crises that demanded some response, from either nature or man.[91] Malthus's *Essay* provides an example of the crisis-response cycle. For Malthus, sudden spikes in death through famine, misery, and war were nature's mechanism for containing excessive population growth. The less moderate the men, the more dramatic was nature's response. The more moderate the men, the less nature ruled his fate.

Twentieth-century demographers would repeat and reconfigure the codes and conventions of imperial masculinity assembled by Malthusian logic to meet the historical conditions of their times and places. With growing confidence in disease control and the Euro-American prohibition of immigration, fertility control became particularly pertinent in the mid-twentieth century, and techniques for understanding the implication of fertility for growth trends became the focus of statistical innovation. Those efforts built on stable population theory and a measurement of the true rate of natural increase, developed in 1925 by Albert Lotka and Louis Dublin, statisticians at the Metropolitan Life Insurance Company.[92] Their innovations corrected for the flaws of annual rates, which appeared to fluctuate haphazardly. As such they were deemed "inadequate measures of the inherent dynamics of a population" because their fluctuations were "influenced heavily by temporary aberrations in the age and sex composition of the population."[93] That is, crude birth and death rates are a ratio of the number of events per 1,000 population. But the risks of giving birth or of dying are not uniform within the population. The risk of dying varies across the human life span; it is greatest among the very young and the very old. Likewise, humans can only reproduce during a segment of their adult lives, and only women are at risk of giving birth. Fluctuating rates could represent changes in the age or sex composition rather than long-term trends in the population. Thus, fertility rates can "go higher without any change in the reproductive behavior of each woman, simply because of the changing age structure."[94] Dublin and Lotka theorized that for any specific fertility and mortality "schedule" a population's underlying age and sex structure would become stable and unchanging over time.[95] Therefore, using actuarial tools—life tables and survivorship curves—it was possible to build mathematical models of the age and sex composition of a population.[96] Age pyramids are the most common form of such models (figure 2.1). These mathematical models made it possible to account for the influence of age and sex on fertility and mortality patterns and reveal a population's "true" rate of natural increase.[97]

"Here were new tools that permitted analysts to plumb beneath surface manifestations," Irene Taeuber noted in 1946.[98] That is, by calibrating the prevailing mortality and fertility patterns by age groups, demographers could calculate a population's net reproduction rate, the likelihood

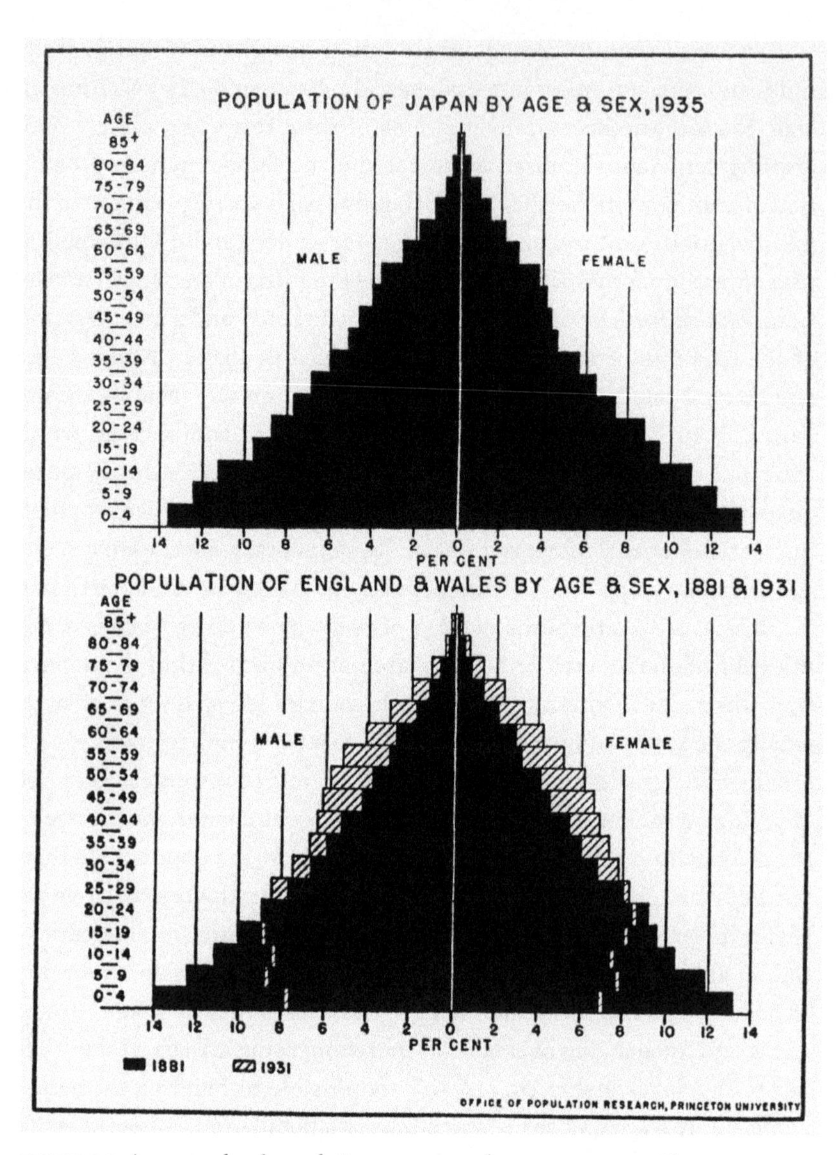

FIGURE 2.1. An example of population represented as age-sex pyramids. From Notestein 1945b, 49. Courtesy of the University of Chicago Press.

that a person would be replaced in the next generation. Reproduction rates thus provide a "measure of the future effect of present demographic behavior," and they are a very common demographic method of representing a population's growth potential.[99] A reproduction rate above 1 means the population is increasing. The reproduction rate calculations do not generate particularly striking figures. But, the most common demographic representation of those rates does produce striking figures. Expressed as doubling time, a reproduction rate of 1.2 will result in a population doubling in size in 57 years. With a 2 rate it will double in 35 years, and with a 3 rate it will double in 23 years.[100]

In practice, net reproduction rates are gendered; they represent the probability that a woman will be replaced by another woman in the next generation. But the ratio is not composed of actual mother-daughter pairs, just women and girl ratios across generations.[101] One could also track men and boys. But paternity was not considered by Dublin and Lotka to be an independent factor; it was "determined" by "assortative marriage" patterns. Maternity, however, is conceptualized as a natural feature of female bodies, so it is women who are routinely tracked in fertility rates.[102]

With the equations of stable population theory, demographers produced new statistical methods that enabled them to register the growth potential within any actual unstable population. Demographers valued these methods because they improved comparability of population statistics. By controlling for differences in age structures between regions, nations, classes, and races, it was possible to assemble a single scale of population dynamics and compare the relative position of regions, nations, races, and so on, based on their vital statistics.[103] Demographic forecasts of future population trends built on these methods. However, such projections are "hypothetical because the vital rates of actual populations are never absolutely constant." Moreover, they are hypothetical because they depend on projection into the future of present circumstances or specific changes in those circumstances.[104] Therefore, they are a theory of the future. They are not a prediction of what will happen, but of what might happen if the assumptions in the model prevail in the world. Nonetheless, through such "chains of observations of observations," demographic knowledge solidified the normative judgment that the world was already overpopulated, that parts of it were suddenly growing too rapidly, and that the effects of current patterns would reverberate into the future.[105] That narrative of population change still prevails and is updated in annual population projections by the United Nations.

Overall, Malthus's *Essay* proved an invaluable touchstone for twentieth-century demographers. It supplied an authoritative ground upon which to expand their claims for the value of statistical accounts of population change for public policy. It supported their claims of impartiality through both the mantle of mathematics and the encrustation of Malthusian hegemonic masculinity in the archives of population history. Rereading Malthus as a masculine narrative of reproductivity illuminates the robust imperial fictions of gender, generation, and nation that organized his mathematics. Demographers took up this Malthusian logic and reconfigured it for the demands of the mid-twentieth century. They deployed their statistical techniques to elaborate the scope of the population problem and establish target rates for increased contraceptive use, reduced fertility, and stabilized growth. Based on the projected growth figures, national population control programs both intervened into reproductive bodies and conducted campaigns to instill contraceptive practices into the local discourses of self-mastery and personal happiness.[106] Through its discourse and practice, mid-twentieth-century demography joined disciplinary and regulatory instrumentalities of governance in population figures that moved nations to act to control population growth through reduction of aggregate fertility rates. In this way, the theory of social reality animated by demographic figures contributed significantly to the dynamic nominalism that produced lower fertility worldwide.

The subject of mid-twentieth-century demographic reason, a (white, bourgeois) liberal American man, was the heir to the (white, bourgeois) rational English gentleman who stood atop the Malthusian scale of civilization.[107] The next chapter takes a closer look at the social worlds of midcentury demography, highlighting the uses demographers made of their quantification practices and gendered geopolitics to mark the boundary of that world, creating a Euro-American male-dominated space in which demographic accounts of fertility were assembled.

3

Narratives of Exclusion, Mechanisms of Inclusion

Demographic Boundary Work

> Sociology must seek also to detect how conditions in the past banished certain individuals, things, or ideas, how circumstances rendered them marginal, excluded, or repressed. . . . Those ghosts tie present subjects to past histories.
>
> —JANICE RADWAY

ON MAY 7, 1931, THREE DOZEN PEOPLE CAME TOGETHER AT THE Town Hall Club in New York City at the invitation of Henry Pratt Fairchild, a leading sociologist, and Margaret Sanger, renowned birth control advocate. US demographers recognize this meeting as the founding event of their national professional organization, the Population Association of America (PAA). The gathering assembled actuaries, census statisticians, sociologists, economists, eugenicists, and birth controllers who were interested in the as yet "somewhat amorphous and unformed" field of population. A confluence of events prompted formation of the new organization. Passage of permanent immigration restriction in the mid-1920s and vast improvements in US birth and death registration systems created new opportunities to analyze the "resident population," especially "fundamental factors of fertility and mortality." At the same time, the International Union for the Scientific Study of Population Problems (IUSSPP) established several years earlier provided a venue for developing comparative demographic measures of "world population."[1]

The nascent science of demography was being nurtured in a number of scattered locations that coalesced in the new national association. The US Census Bureau, which was established as a permanent bureau in 1913, focused on enumeration techniques. At the Metropolitan Life Insurance Company, Louis Dublin and Alfred Lotka laid the mathematical groundwork for demographic analysis of mortality, fertility, and the true rate of natural increase. Warren Thompson, who held the first academic post in population science, and his protégé, Pascal Whelpton, pioneered population projections at the Scripps Foundation for Research in Population Problems at Ohio's Miami University.[2] At Johns Hopkins University, Raymond Pearl, who earlier had studied reproduction in fruit flies and chickens, pursued the biometrics of human fertility patterns.[3] Guy Irving Burch set up the Population Reference Bureau at New York University in 1929 to disseminate population information for policymakers.[4] The Milbank Memorial Fund (MMF), established in 1905 to promote public health policy solutions, added a population division to its operations in 1928. The MMF, which provided financial support for the founding meeting, would contribute critical support for the discipline's development into the 1950s.[5] Margaret Sanger and Eleanor Jones, president of the American Birth Control League, were among the few women in attendance. They represented the so-called applied side of population science, the birth control clinics in which research on contraception and its use would be carried out during the coming decades.

The events of the May 7, 1931, meeting have been recounted so often and in so many disciplinary venues that they have become part of the discipline's folklore. The version quoted below was presented during a special commemorative panel at the association's fiftieth anniversary in 1981.[6] Disciplinary histories also encode this narrative into the official story of the founding of the field, and it has become a staple feature of general historiography on the twentieth-century population control establishment.[7] One pivotal moment of the founding story stands out as fundamental to the organization's destiny. That is, in the midst of electing officers, "an embarrassing situation developed."

> It was Dr. Fairchild's idea that Mrs. Sanger should be the first president. He nominated her, and there was a long silence. There were no other nominations, and Fairchild was about to ask for the vote. The atmo-

> sphere in the room seemed tense. Finally, one of those present rose to suggest that though Mrs. Sanger was one of the great women of her time, her election to the presidency of the Association would fail to confirm its scientific image. The purpose of the scientists who were attending the meeting, and indeed the purpose of Mrs. Sanger herself, was that the PAA should be a group of scientists solely interested in research and publication. As such and only as such could it usefully serve as the scientific foundation for birth control activities. Mrs. Sanger at once rose and graciously agreed. Tension was lifted, and Professor Henry Pratt Fairchild was elected the first President.[8]

This master narrative vividly illustrates the disciplinary common sense: the field was founded in an act of rescue. The person who spoke up "saved the association."[9] By ensuring that it would be separate from Mrs. Sanger and birth control, he ensured that it would attain its rightful stature as a scientific organization. The story encodes the strongly held belief that the field is uniquely vulnerable to outside political influences and has had repeatedly to distance itself from such influence in order to safeguard its scientific reputation. From this perspective, the story and its ritual repetitions make sense of demography's past, define its community in the present, and (re)affirm the proper path for its future.[10]

I retell the story for a different purpose: to unsettle the conventional wisdom it inscribes. Using the tools of feminist science studies to read against the grain of disciplinary histories,[11] this chapter illuminates what the plot and characters of the master narrative convey and conceal about the gendered power relations that configured the social worlds of mid-twentieth-century US demography.[12] The chapter traces the exclusions (and inclusions) of persons and perspectives within the social landscape that would constitute population science. It argues that Sanger's exclusion was a crucial instance of disciplinary boundary work by which demographers established and reaffirmed their cultural authority over population matters.[13] The act of drawing boundaries configures an authoritative community around particular practices and epistemologies and thus is central to disciplinary formation. Boundary work sets limits upon the "disparate elements" that can be "brought into particular types of knowledge relations" by a discipline's "objects of study."[14] In particular, boundary work consists of the "exclusion of those offering discrepant and competitive" viewpoints in the political and "intellectual landscape." When successful, disciplinary boundaries bracket credibility and lend cultural authority to knowledge claims, insulating them from questions about what interests and motives drive them. Thus to understand the boundaries of any particular science, one must investigate the particular history of "when, how, and to what ends the boundaries of science are drawn and defended."[15]

From this perspective, the most interesting aspect of the story is that it "vividly" marks the boundaries of the field through a gendered act of exclusion and then represents this action as fundamental to credible science making.[16] The folklore of the PAA's founding, I contend, both enacts and exemplifies the repeated gestures by which demographers staked their claim to scientific credibility to their distance from the woman-led birth control movement. Such gender-inflected gestures became a reflexive mechanism by which demographers defined and distinguished the field for themselves, their scientific colleagues, their competitors, policymakers, and the wider public. Moreover, I argue, these gender-inflected gestures became a shield against controversy, helping to establish the discipline as the disinterested arbiter of controversy rather than as party to it. Over time, statements distancing population science from the birth control movement became a ritualized means through which demographers reaffirmed their self-representation as the voice of moderation and the appropriately impartial authority on fertility matters.

The analysis also troubles the conventional demographic narrative by looking within the boundaries of the discipline at what is left unsaid in the founding story, elucidating what was included within the mainstream of the discipline with little comment or controversy. The other two characters in the founding tale, Henry Pratt Fairchild and Frederick Osborn, the latter of whom urged Sanger to withdraw, were actively engaged in eugenics. One historian has called Fairchild the most prominent academic racist of the 1930s.[17] Less well known, Frederick Osborn was a leading figure in the American Eugenics Society at the time of the PAA founding meeting and throughout his life.[18] Contemporary demographic histories discount this connection, suggesting that eugenicists folded into demographic circles by default, but that demographers quickly marginalized them.[19] To the contrary, I argue, eugenicists were crucial to "standardizing the tools of demographic measurement," which became the association's chief objective.[20] But more than mathematics bound demographers and eugenics. Hegemonic masculinity also mattered as both groups sought to understand (and control) the connections between individual reproductive behavior and "the dynamics of group behavior."[21] A close reading of Frederick Osborn's writings elucidates the cultural differentialism inscribed in Malthusian masculinity that explains the mutual unease demographers and eugenicists harbored about the birth control movement's efforts to secure individual reproductive rights for women.

Finally, the chapter investigates the contours of the relationship between demographers and birth control advocates at their margins after demography "captured authority" over population problems and solutions. It demonstrates that although that relationship may have been reciprocal, the two groups did not command equal cultural authority. Demographers exercised considerable authority for assessing the birth control movement's goals, rhetoric, agenda, and clinical practices. Yet as Adele Clarke has noted with regard to the reproductive life sciences, this is a "complicated" story. It involves "negotiations and trade-offs rather than simpler sagas of repression and denial."[22] Birth control advocates deferred to the claims of science. They became a principal audience for demographic knowledge products. In fact, they eagerly and at times incautiously used the certified knowledge created by population scientists to bolster their cause without paying sufficient heed to the sticky realignments between individual and aggregate in demographic accounts. Taken together, the sections of the chapter illuminate the intertwining politics of gender differentiation and hierarchy that structured the social worlds of population science at its foundation and set up a troubled relationship between the "gentlemen's club" of demography and birth control advocacy.[23]

Narratives of Exclusion: Margaret Sanger and Birth Control

How did the fiftieth anniversary commemorative panel quoted above account for the tension and embarrassment caused by the possibility of Sanger's election to a leadership position in the PAA? Why would her participation threaten the organization's future success? It is not as if Sanger was an interloper at the initial meeting. To the contrary, Sanger and Fairchild had jointly issued the invitation to attendees. Moreover, Sanger had secured the financial backing for the meeting from the Milbank Memorial Fund, and her staff provided logistical support.[24] In addition, the panelists—Anders Lunde (then PAA historian), Frank Notestein, Frank Lorimer, and Clyde Kiser, all of whom had attended the founding meeting as young scholars—observed that opposition to birth control did not motivate Frederick Osborn's intervention. They insisted that he admired "Mrs. Sanger" and "made regular contributions" to the birth control movement. The issue at stake was control of the organization and its agenda. Osborn's intervention addressed the "undercurrent of concern lest birth control advocates dominate the association."[25] The panelists pointed to the standards of scientific professionalism to account for why it was necessary to exclude Sanger from the association's leadership and to exclude birth control from the organization's initial research agenda. They noted that Osborn "spoke eloquently about the dangers" of pursuing any reform, which would bring "emotional elements" into the organization that might damage its chances of success as well as its reputation. He spoke to the scientists' "responsibility of guarding against religious, moral, and political dogmatism in the area of population."[26] With this logic, Osborn "persuaded the meeting that the fortunes of the field would be advanced if the new Association were to guard its scientific nature and keep free from attachment to the birth control movement."[27]

Such statements signify demographers' adherence to the conventional distinction between science and advocacy, or "propaganda," as they labeled it. They deployed the cultural authority of science—its "voices of calm and moderation," its impartiality, its avoidance of controversy in favor of basic research—to legitimate the exclusion of Sanger and birth control.[28] As Frank Notestein framed the matter, the job of science "was to discover not to preach." Scientists' interests "lay in knowing not doing." The main problem was that "the birth controllers . . . were willing to beat the drum with any stick that came handy." Science was simply "a means of rallying support, but the object of the [birth control] enterprise was the reform itself." Therefore, founding members of the discipline argued, to become a credible scientific organization, the PAA had to keep its distance from birth control. It had to take a different, more measured, tone. In that way, objective research would lead the way, not propaganda. Once initial research findings set the course, Notestein argued, that knowledge would produce its own momentum. Each new finding would dispassionately suggest the next appropriate research question.[29]

Retellings of the story at commemorative occasions reiterate these disciplinary commitments. Told in retrospect, the events are narrativized, characters are fixed, and the outcome becomes certain. Retelling the story performs a pedagogical function as well. It reminds the audience of students and practitioners what falls within the proper scope of demographic practice. It reiterates who and what belong within the discipline and who and what does not belong. The retellings are thus examples of "institutionalization in motion"; they continually re-fix the margins of the field.[30] In this regard, it is noteworthy that the commemorative retellings of this tale of humble but heroic beginnings increased in the 1970s and 1980s, a period when the field came under renewed challenge from feminist reproductive rights activists worldwide.

It is not just at commemorative occasions that demographers stake the boundaries of the field to the exclusion of Sanger and birth control. Formal histories of the field provide similar accounts. One of the most widely disseminated, written by Rupert Vance, appeared in *The Study of Population*. The 1959 publication, edited by Philip Hauser and Otis Dudley Duncan, became "a standard reference for generations of students."[31] Funded by the National Science Foundation as part of its survey of the state of science in America, the volume appeared as the field was becoming ascendant. It summarized the intellectual debates, methodological achievements, and disciplinary allies of the first generation of US demographers.[32] Vance situated the history of demography in the United States within the mathematical lineage of Quetelet and Malthus. Near the end of his account, in a section entitled "Circumstances Affecting the Future Development of Demography," Vance discussed the 1931 founding of the PAA, noting that the lateness of its organization could be traced in part to "lay interest in the controversial aspects of population." He described the early years of the PAA as follows: "In the popular mind, the scientific study of population was still confused with the spread of propaganda for birth control. The organization of the association was delayed in the hope of avoiding entanglement in demographic controversy. . . . Propagandists were not made welcome, and the original organization was carefully set up with only scientists on the executive board. . . . While this choice may have retarded popular education in demographic matters, it definitely has improved the scientific status of demography in the United States."[33] This formal history used the same devices as the colloquial versions of the origin story to mark the boundary of the discipline and thereby establish its credibility. Again, the contrast was between propaganda and science. Again, birth control represented the controversy that the field had to avoid. To police these boundaries and its reputation, the association, Vance noted, put in place a two-tiered membership structure with the governing College of Fellows formally separated from the general membership.[34] This group appointed the association's board, which in turn defined the association's rules and elected its officers. In this way, a small group of closely allied men controlled the organization and its agenda. This structure remained in place until 1954, when the professional and scientific legitimacy of demography stabilized and resources flowed steadily to support its emerging institutions.

Vance presented the history of the early PAA in a narrative that otherwise emphasized the discipline's mathematical sophistication. In his account, the demographic body of knowledge developed through the progressive discovery of mathematical laws of population, which had nothing to do with politics. The scientific basis of demography is thus represented as independent of this organizational history in the same way that all mathematical truth is taken to be independent of the social context of calculation. The numbers are what the numbers are. They are to be discovered through the rigorous procedures of experts. Problems were represented as occurring when "laymen" and propagandists latched on to the numbers to bolster their self-interested claims. Thus Vance praised the association's early structure, concluding that steps taken to freeze out propagandists may have been undemocratic, but they were necessary to protect the scientific enterprise from sensationalism and controversy. Both the popular and disciplinary histories thus valorize the boundary work performed by the founders as necessary to protect the autonomy and integrity of the discipline.

However, the distinction demographers made between their science and birth control politics, between investigation and action, between the disciplined and the unruly, is itself a political gesture. Demographers are not unique in relying on these binary distinctions in an effort to establish their scientific credibility. They made use of a common device for establishing the credibility of positivist scientific disciplines: setting the boundaries of what counts as science is a key component of securing credibility for particular practices. As Thomas Gieryn notes, "Maps of science get drawn by knowledge makers hoping to have their claims accepted as valid and influential downstream, their practices esteemed and supported financially, their culture sustained as the home of objectivity, reason, truth, or utility."[35] Drawing a boundary between themselves and birth control advocates, demographers displayed the field's conformity with and commitment to the standards of scientific impartiality. By investing in the hallmarks of science—quantification, universal laws, disinterested inquiry, and moderation—population scientists claimed objectivity and intellectual rigor for themselves.[36] The mantle of science secured cultural authority for the demographic map of the world.

Standardization of demographic statistical techniques in the 1930s facilitated the demographers' capture of epistemic authority for fertility control, certainly. Yet the founding events defined the in discipline's boundaries at a moment before consensus had even formed around demography as the name for the new discipline. This consensus emerged as the scope, objects, and standard methods of the field were beginning to be articulated. The PAA's distancing itself from sentiment and controversy through the exclusion of Sanger, therefore, cleared the initial ground on which the US community of population science coalesced.[37] Far from being merely an amusing anecdote in the history of the discipline's intellectual achievements, this exclusion constituted the first definitive act that established the discipline. That is, before statistical tools had been validated, before research results had been amassed, before the disciplinary name "demography" had been settled upon, the act of excluding "Mrs. Sanger" and "birth controllers" from leadership roles provided a powerful demonstration of the discipline's scientific mettle. It guaranteed the grounds on which the association would tout itself in public discourse as an organization in which impartial "technical work" could dispatch controversy.[38]

The contrast between demographers and birth control strengthened demographers' self-representation as impartial, objective, and moderate. That is, the epistemic authority of science depends on designating and distinguishing one's claims and practices from those "unworthy claims and practices of some nether region of *non*-science."[39] By affixing the label of nonscientific propaganda to birth control advocacy, demographers configured and protected their epistemic authority through gestures that simultaneously denied that authority to birth control advocates. Yet these exclusionary gestures simultaneously bound birth control advocacy to demography's margins as the "abject" other against which its credibility and authority were calibrated.[40] This bit of boundary work effectively defined population science as inside the field of demography and birth control advocacy as outside.[41] The exclusion of birth control advocates from the field of science denied their agency—Sanger's only act is to withdraw—but reintroduced them as abject objects whose noise on the margins defined demographic moderation. As Frank Notestein observed, birth control advocates were drumbeaters. Ironically, while the first generation of demographers could "scarcely conceal their contempt for propaganda that advocated birth control," the birth controllers, those "interfering feminists" at the margins of the discipline, were nonetheless indispensable.[42] In a sense, then, demographers had an investment in the continuing binary representation of their science in contrast to birth control politics.

The distinction demographers made between their science and birth control propaganda worked as well as it did in part because it aligned with conventional representations of gender differences. As Thomas Gieryn notes, boundary work involves the "attribution of selected characteristics" in order to construct a "social boundary" that demarcates the space of science.[43] Boundaries make a difference. In fact, as Adele Clarke notes, "all boundaries are about difference claims."[44] The public activities of population scientists on policy issues overlapped considerably with those of the lay propagandists. Certainly, they both sought means to control birth rates. However, one difference stood out: their gender. The birth control advocates were mostly women and the population scientists were overwhelmingly men. In a historical period in which reason and emotion were defined as distinctly masculine and feminine traits, respectively, gender served as an obvious characteristic to mark the social boundary of demographic science.[45] While the men could move easily between advocacy groups and scientific organizations, the women could not. They did not possess the requisite credentials. Men could speak as citizens and scientists; not so women. Given the restricted access to the halls of science, academic credentials were also badges of masculine privilege that demographers leveraged to portray birth controllers as inexpert laywomen.

Osborn relied on normative conventions of gender difference to enunciate the differences between demography's scientific credibility and birth control politics when he observed that although "Mrs. Sanger is one of the great women of her time," her presence as a leader would "fail" to symbolize the scientific nature of the organization's expertise. She may be a great woman, but she cannot symbolize science. The wifely mode of address so common in demographic narratives amplified this distinction of rank even further.[46] As "Mrs. Sanger," she was merely a lay advocate, simply a woman reformer, a creature whose position in the public world of politics was unsettled and unsettling.[47] Notestein's characterization of birth controllers as "drumbeaters" explicitly recalls the spectacle of feminist agitation, of suffrage marches, in contrast to the moderate discourse of science. With this exclusionary act, not just troublesome women were banished from the field, but women's gendered perspectives on reproductive issues were also pushed to the margins, outside of the legitimate domain of population science. As Notestein identified it, the drum Sanger beat was "the cause of women's right to control their reproductive destiny."[48]

The founding of the PAA was not the first time gender marked the boundary of population science and propaganda. The same gendered politics of expertise shaped an earlier meeting of population scientists brought together by Sanger. In 1927, Sanger and her associates organized the Sixth International Malthusian Conference in Geneva. This meeting, which brought together population scientists from across the globe, laid the groundwork for the establishment of the International Union for the Scientific Study of Population Problems (IUSSPP) the following year.[49] Sanger and a small group of women volunteers secured funding for the conference and managed all the logistical tasks for the event. However, when the time approached, they were muscled to the margins. The IUSSPP founding is often also mentioned in accounts of the PAA founding; US demographers attribute Sanger's exclusion to Raymond Pearl, the advisory committee co-chair with Sir Bernard Mallet.[50] Pearl was involved, but Sanger's 1938 autobiography recalls that it was Mallet who pushed the women aside. On the eve of the conference, he removed the women's names from the official program. Sanger recorded that when she objected, Mallet declared, "The names of workers should not be included on scientific programmes." Sanger countered that "in their particular lines they are as much experts as the scientists." Sanger insisted "that in justice to the women who had given so generously of their time and effort, who had raised the money, issued the invitations, paid the delegates' expenses, they should be given proper credit." However, giving women credit was precisely the problem. As Sanger was later informed, Mallet had been warned that "these distinguished scientists would be the laughing stock of all of Europe if it was known that a woman had brought them together." In the end, Mallet prevailed, the women's names were removed, and the reputation of the scientists was preserved.[51] Women's perspectives on reproductive processes were also excluded through the embargo on discussion of birth control.

The anxiety about the presence of the women expressed by participants at the 1927 and 1931 founding conferences speaks to the still unsettled cultural authority of social scientists. In the first years of the twentieth century, several major universities fired prominent social scientists who stood up for the perspectives of labor activists.[52] In response, the social sciences increasingly relied on scientism to secure autonomy in the academy and establish their cultural authority. Social scientists declared their commitments to positivist practices of quantification, objectification, and value-neutrality in service to building systematic knowledge.[53] Such declarations demonstrated the commonality between the methods and metaphors of social and natural sciences, enabling emerging social science disciplines to claim a degree of universalism that insulated them from political scrutiny. This universalism, "the search for basic principles" underlying the "general laws" of "human life," relied on the distinction they made between "pure" and "applied" research. The distance provided by pure research, carried out "at a distance" from any specific context, removed the influence of particular interests.[54] Applied research, conducted within specific contexts, on the other hand, was mired in the particular. Thus, "pure research" was necessary to "overcome the obstacles of prejudice and bias."[55] The Rockefeller-sponsored Social Science Research Council, whose funding practices set the social science research agenda of the period, rewarded the intellectual investment in "value-neutral" science and those who distanced themselves personally from the politics of troublemaking.[56] By the 1930s, as Charles Cooley noted of sociology, "Our objective now is to understand rather than to do."[57]

The politics of troublemaking were gendered, as represented by women's demands for a greater voice in public life. Feminist sociologists have noted that the scientism that their discipline advocated in the interwar period was deeply masculinist. It drew heavily on existing metaphors of white bourgeois masculinity—rationality, dispassion, and intellectualism—to bolster professional claims.[58] Women, who had only recently obtained access to higher education, still had a very tentative position within the modern university.[59] But women did establish influential positions within local communities, from which they entered into social science debates. The discourse of pure versus applied research that gained purchase in this period reaffirmed the subordinate position of women as public actors and knowledge builders. The dominant sociological "school" of thought in the 1920s, the Chicago School, led by Robert Park and Edwin Burgess, touted the science of sociology in what would become the first authoritative textbook for the field.[60] They worked with the Social Science Research Council to shape the research agenda of the twenties and thirties and at the same time worked assiduously to distance their theory and practice from the practical empirical research of social reformers. In particular, they dissolved their longstanding connections to Hull House's women social scientists. At the same time, scientism refashioned older cultural claims to the superiority of the perspectives of bourgeois white men over a host of others. By underwriting the claim to value-neutral knowledge with the metaphors of hegemonic masculinity, scientism enabled the continuing hegemony of white bourgeois men for definitions of the common good, including gender norms. Sociologists' investments in scientism, as a response to women's challenges to prevailing gender conventions, were "an attempt to modernize middle-class white masculinity."[61] Of course, scientism's masculine modernization was invested in hierarchies of masculinity as well, which is discussed below.

The gendering of other social sciences in the 1930s was similar. The emerging field of population studies was yet one more social science that staked its professional credibility and autonomy on masculine scientism. Doing so, demographers deployed a powerful tool. With the authority of science, they would construct the statistical tools to discover "the facts," the general laws that governed the relationship between individual reproductive performance and aggregate population trends. Statistics, mapping, and related numeric representational practices provided especially strong evidence of objectivity because numbers are solid, fixed, certain. They do not change with the context. Paired with the gestures distancing themselves from birth control, the claim that demographers were only interested in mathematical truth resonated more strongly. Birth control advocacy, written in the discourse of feminine sentimentalism, provided ample evidence with which population scientists could build the contrast to their scientism. Nonetheless, the success of this contrast rested in the sticky gender politics underlying the pure-applied, science-action binaries themselves.

Once birth control advocates had been excluded from the social world of demography, it became easier to typify them as troublesome amateurs and propagandists.[62] The distinctions of science and action reverberated with the metaphors of hegemonic masculinity to amplify the bias attributed to birth control advocates. Later colloquial accounts often include explicitly disparaging comments about the intellectual caliber of birth control advocates, noting their lack of a proper understanding of the determinants of fertility. As Frank Lorimer's 1971 reminiscence notes, Sanger "was not severely constrained in her thinking by scrupulous respect for the findings of scientific inquiry."[63] Even as recently as 1995, one scholar noted in a typical example that birth control advocates were "unrestrained by the necessity to maintain scholarly consistency."[64] In later years, Sanger, birth control, and "planned parenthood nuts" would become figures of derision.[65] The truth of their negative evaluation of birth controllers is self-evident to demographic commentators and conveys the message that arguments and analyses of birth controllers need not be accorded the same respect as those made by demographers.

The value of this binary contrast with birth control propagandists for "the fortunes of the field" was assessed in the 1982 disciplinary history Frank Notestein wrote at the end of his long and distinguished career. Reflecting on the PAA strategy of pursuing "basic research," he argued:

> Probably the best way to make progress in a dangerous field is to sponsor "research" rather than "action." Who can be against "truth"? Study the field, and let everyone come to his own conclusions, the slogan can be. Keep clear of advocacy. So research becomes a substitute for action. It is precisely this tactic that has helped foster research in the demographic field. With all its delay, I am inclined to think that it is also

> the best way to foster "action." In a sense, "dangerous" subjects have attracted research funds precisely because they prepared the way for, and permitted delay on, the action front.[66]

Notestein observed that demography, "dealing as it does with birth, death, marriage, divorce . . . the very stuff of life, is peculiarly sensitive." Although an issue for all social science research, demography is particularly vulnerable because it "impinges on deeply held values."[67] Although sensitive, such research was "indispensable" because, he noted, "if a subject is sensitive, there is always a lunatic fringe. But lunacy does not spawn technical work." Research "supplies the essential array of controlled information and inference and underlies the responsible formulation of public policy in a rapidly changing world."[68] The claim to objective scientific truth in contrast to the unruly birth control movement is said to have assuaged the hesitation of foundations to finance "sensitive research" on reproductive matters.[69]

Statistical forms of knowledge also had another benefit, which Notestein identified. With them, demographers could investigate human reproductive processes without "the taint of human sexual interaction." As Adele Clarke has argued with regard to the reproductive life sciences, the connection to human sexual practices made these disciplines risky endeavors that scientists negotiated by using the basic/applied research dichotomy to distance their work from sex acts.[70] Population science, "the mathematics of the process of self-renewal," avoided the sexual details of reproductive performances, decreasing the risqué connotations of population knowledge building.[71] As detailed in the next chapter, the mathematics of self-renewal also elides the gender power relations that comprise heteronormative sexual/reproductive practices. Here, it is important to note that the modernization of white, middle-class masculinity accomplished through population science reaffirmed the subordination of women in public and private. It also recaptured the terrain of reproductive self-determination for its vision of the national interest: The "single-minded" focus on advocating women's individual rights represented the unreasonable bias of birth control advocates, which did not take adequate account of population dynamics in general.[72] Demographers thought birth control advocates did not appreciate the impact individual behavior could have on national and international population trends, which was the primary concern of demographers. Leaving the vitality of the national population to the conscience of individual women was not, in their view, sound public policy. The "value-neutral approach" articulated by demographic science, on the other hand, was balanced. With statistical tools, they would be able to identify the population "danger spots" around the globe that resulted from ill-considered and irresponsible actions of individuals, and would be able to direct public officials to effective solutions for managing aggregate population dynamics in the national interests.[73]

The gender politics at work in the distinction demographers drew between population science and birth control politics is even more striking when one compares demographers' portrayal of the careers of Margaret Sanger and Frederick Osborn, the hero of the founding tale. Like Sanger, Osborn was not a credentialed scientist. He was a son of wealth, who, having had a lucrative career in business, "decided to devote his energies to something he liked."[74] As demographers would later recount, in 1928 he prevailed upon his famous uncle, Henry Fairfield Osborn, the renowned eugenicist and curator of the New York Museum of Natural History, to give him office space and the title of research associate. With these, he pursued an independent investigation in the fields of biology, eugenics, and population, topics that had interested him for many years. After two years of study, he dedicated himself to the task of promoting population science.[75] Osborn was welcomed by the "gentlemen's club" of demography. In fact, the gesture that excluded Sanger represented the initiation of Frederick Osborn into the boundary work of demography. He was elected to the PAA executive committee at the founding meeting.[76] He went on to become very influential in the field, serving as president of both the PAA and the Population Council, one of the most important international demographic organizations in the midcentury. His lack of scientific credentials, therefore, was not a disqualification.[77] At the 1981 commemorative panel, Clyde Kiser identified Osborn as "revered" by demographers "throughout the world" "for his lifelong promotion of scientific studies of population."[78] In particular, as Frank Notestein declared, "There is no man to whom the Association owes more for its early tone as a scientific society than Frederick Osborn." Nor, Notestein said, had anyone "done as much to bring major resources to the scientific study of population."[79]

Osborn's position as a man of wealth helped him immensely in his efforts to bring resources to demography. He counted John D. Rockefeller III and Albert Milbank, head of the Milbank Memorial Fund, among his friends.[80] Beyond his access to philanthropic resources, Osborn shared the hegemonic masculinity that defined the social worlds of demography.[81] In subsequent years, no one would work harder to police the boundaries of demographic science; and no one did more to promote the binary imagery of demographers as quiet, moderate scientists and birth control advocates as noisy, reckless propagandists. As discussed below, Osborn's correspondence, his published writings, and the records of the Population Council are replete with instances of the distancing gesture enacted as part of the claim to credibility for demography and its institutions. Osborn repeatedly insisted that he preferred a quiet approach, relying on scientific credibility to influence the wider political and moral debate.[82] Those gestures regularly relied on normative gender differences to reaffirm the line between population science and birth control propaganda.

Frederick Osborn and Eugenics

The boundary work Frederick Osborn performed concerned more than establishing manly population science in contrast to womanly birth control politics. He was also instrumental in shaping the neo-Malthusian hierarchy of masculinities, a hierarchy that sutured together eugenic logics and demographic scientism. By focusing on the risks posed by birth control advocates, disciplinary histories obscure the eugenic concerns invested in their discipline at its founding. As Clyde Kiser noted in his presentation for the 1981 panel, the initial structure of the PAA was designed "for protection against actionists in birth control not eugenists."[83] Throughout the thirties, the membership of the PAA and its governing body overlapped with that of the American Eugenics Society and included eugenicists and nonscientists like Osborn.[84] Eugenic goals were written into the association's mission. The organization's charter included a commitment to "the improvement, advancement, and progress of the human race by means of research with respect to problems connected with human population in its quantitative and qualitative aspects."[85] Demographers also published articles in eugenic journals such as the *Eugenical News* and the *Journal of Heredity* into the 1960s.

Population scientists mingled easily with eugenicists, with whom they shared a commitment to Malthusian logic, statistical forms of knowledge, and a gendered coloniality of power that aligned with their anxieties about the future. Both were concerned with the superabundant effects of reproduction and the impact of individual reproductive performance on differential growth and national vitality.[86] Both worried about the impact women's individual actions would have on those rates. And both invested in statistical forms of knowledge as the means to understand and manage reproductive processes in the nation's interests in a hierarchical world of men and peoples.

Yet, in the 1930s, the complex relationship between individual and group fertility patterns was undertheorized, and that deficiency shaped the mutual commitments of demographers and eugenicists to develop robust statistical tools with which to determine "the facts." The relationship between individual and group processes also informed their shared misgivings about the goals of the birth control movement. Demographers and eugenicists both frowned on the movement's emphasis on limiting fertility, arguing that population control policies had to give proper regard to society's interests in population growth. In the Global North, some large families were needed, they argued, to offset the impact of infertility and mortality on aggregate growth.[87] There was a specific gender focus to these concerns. Echoing race suicide rhetoric from the turn of the twentieth century, some, such as Louis Dublin and Warren Thompson, "deplored" the trend toward increased "equality for women" in education and occupations because it would discourage the "exceptionally endowed" from having large families.[88] The "better stocks" of women were already having "too few children," leaving the task to those less suited to producing high-quality children. Thus eugenicists easily melded into demographic social worlds through their common interest in statistical forms of knowledge along with mutual concern about the impact of women's growing autonomy and of differential fertility on the hierarchy of men and nations.[89]

Of course, in 1931, eugenics still carried some of the imprimatur of science, and many in the scientific community assumed that, at least in part, population dynamics derived from biological processes.[90] This would change over the course of the decade as scholars debated the relative weight of biology and environment in human development. Among demographers, this shift played out in a local version of the period's nature/nurture controversy, encapsulated in the definitional debate over the distinction between fecundity and fertility; this debate is a local version of the quintessential natureculture configuration identified by Donna Haraway.[91] Demographers recount this history through the career of Raymond Pearl, a prominent population biologist and eugenicist.[92] In brief, throughout the 1920s, Pearl was the principal champion of Karl Pearson's biometrics in the United States. He spent the early thirties promoting the biologically based s-curve model of population growth, which he believed described the basis of population growth and decline that both Malthus and Darwin had described. His model rested on the concept of fecundity, the biological basis of prolificacy, as the central tenet by which demographers should describe the processes underlying birth rates. Pearl assumed that differential fertility resulted simply from biologically based class and race differences in reproductive capacity. Others, particularly Frank Notestein, argued that birth rates measured the performance of reproductive capacity, which resulted from the influence of yet unidentified social and psychological factors. By the end of the decade, Pearl's own research would demonstrate that differential access to contraception accounted for differential fertility rates. He then conceded that social factors shape birth rates.[93] Pearl's concession marks the coalescence of consensus on terms. The disciplinary meaning of fecundity was settled as the biological capacity to reproduce, while fertility was defined as its social performance, and birth rates were said to measure that performance.[94] The standardization of this terminology indexes the ascendance of a sociological conceptualization of aggregate population trends because performances take place within a social context. With this definitional resolution, demographers focused on culture and its psychological impacts as the cause of class, race, and national differences in fertility patterns.

The epistemological shift to biosocial explanations of fertility made competent research on the social determinants of reproductive performance all the more urgent. Osborn worked assiduously to promote pioneering research into "the connections between individual reproductive behavior and the dynamics of group behavior."[95] He also promoted a reformed eugenics that emphasized social as well as genetic determinants of character and social worth.[96] For this reason, as Ramsden notes, the PAA "welcomed Osborn's conception of eugenics," and insider histories of demography credit Osborn with "rescu[ing] eugenics from its racist and social class bias" and developing it "on a scientific foundation."[97] His writings therefore provide a critical vantage point on the eugenic perspective that circulated within the social worlds of demography during the period in which demography configured the population explosion as its object of study and articulated population control principles to manage it.[98] A close reading of Osborn's writings demonstrates, however, that far from disengaging eugenics, the sociological turn in population science expanded the reach of eugenic logic.[99] His writing offers an early and rich example of cultural differentialist thinking about race and coloniality.

Osborn's first book, *Dynamics of Population: Social and Biological Significance of Changing Birth Rates in the United States,* published in 1934 with Frank Lorimer, reassessed the significance of differential fertility for aggregate population trends in the United States.[100] Their influential analysis both drew on and shaped demographic thought about the balance of economic, social, and genetic causes and consequences of differential fertility through World War II.[101] Osborn wrote two other books for general audiences, *Preface to Eugenics,* in 1940, and *Population: An International Dilemma,* in 1958. *Preface to Eugenics* reiterated the main arguments of the 1934 book, giving fuller attention to questions of race and class difference, as well as the social control of population. *Population,* as the title indicates, extended his arguments about the problem of differential fertility to an international frame. It grew out of a two-year study group he sponsored through the Population Council that was designed to review the best thinking on the topic. The book laid out the agenda of the Population Council, making the case for systematic data gathering and state-sponsored fertility reduction efforts.[102] All of his books highlighted the need for additional scientific research to guide reasonable and effective social control of individual reproductive practices to optimize group trends.[103]

Throughout his writings, Osborn specifically disavows earlier eugenic claims about racial differences, arguing that they derived from irrational prejudice and emotional bias, not science. In *Dynamics of Population,* Lorimer and Osborn contested eugenicist interpretations of IQ scores on tests administered to US Army recruits during World War I. Reanalyzing the data themselves, they concluded there was no significant scientific evidence of innate racial differences. Their new statistical analysis instead showed that "wide differences in the cultural-intellectual development of the different races . . . are in every case associated with variations in occupational status and in regional location," noting in particular that African Americans from northern states scored as well as southern whites. They concluded, "It is unsound to deduce . . . that race is a cause of group differences in intelligence." They further argued that "indirect evidence is sufficient to prove conclusively that the apparent differences in cultural-intellectual development between major racial groups are due in large part to environmental influences." Osborn did not go so far as to conclude that there were no biological differences, only that so far the data were ambiguous and therefore that it was "wise absolutely to reserve judgment about the comparative hereditary capacities of large racial groups."[104]

In *Preface to Eugenics,* Osborn reiterated, "There is as yet no scientific evidence as to whether these races [African Americans, Mexicans, and Indians] differ from the white stock in genetic capacity to develop qualities of social value." Moreover, he equated race with class, asserting that there was no evidence that racial differences "are more influenced by hereditary factors than are occupational differences." He argued to the contrary that the great diversity within large social and racial groups meant that one's membership in a particular group alone did not determine one's genetic endowment or social worth. Instead, he estimated, 40% of those in the lower occupational classes actually had "hereditary capacities above the average of the upper groups." With regard to race, Osborn did not offer a precise estimate, noting that "almost nothing is known as to the relative hereditary capacities" of nonwhite racial groups. But, he asserted, "their lower environmental level must act as a serious deterrent to the upward development of their intelligence."[105] Thus, as a preface to eugenics, environmental factors should be equalized. Because large social groups contained individuals of varying capacities, Osborn concluded, an approximately equal environment of health care and education would allow those with stronger genetic endowment within each group to rise to the top. But, he also noted, once the environment was approximately the same for all groups, "there would still be individual differences in intelligence, and heredity would be their sole cause."[106]

On the surface, Osborn's rhetoric about innate racial differences appears to be moderate, as demographers claim. However, Osborn brings the eugenic philosophy of better breeding back in through cultural differentialism. That is, as Étienne Balibar has argued, "While moving from . . . the language of biology into the discourses of culture and history" after the demise of eugenics, mid-twentieth-century racial discourse "retain[ed] the same structure." Lorimer and Osborn exemplify the continuity Balibar describes. They both conceded a measure of equality among human groups by accepting environmental determinants of difference; they then recuperated the conventional hierarchy among those groups through "a logic of a naturalization and racization of the social." As Balibar notes in the grammar of cultural differentialist theories of difference, "The signifier culture" is "always . . . attach[ed] to heritage, ancestry, a rootedness" that rearticulates "man" to "his origins."[107] For Osborn and Lorimer, "man's social heritage" became the primary source of environmental determinants of character, accomplishment, and social worth. Thus, they asserted, "the cultural level of each generation determines in a high degree the environmental factors that affect intellectual development of the next." The exact terminology was not yet settled in 1934, but as they defined it, cultural level consisted of the modes, standards, and planes of living that were shaped by heritage, habits, and traditions of the family and community in which one was raised.[108]

Further research was required, Osborn contended, before reasonable and valid conclusions could be drawn about the exact nature and extent of qualitative differences between large social and racial groups produced by both cultural level and genetic endowment. In the meantime, he argued, some conclusions could be drawn about the impact of qualitative differences among individuals and large social groups. First, he noted that the pattern of early childhood IQ test scores was positively correlated with later academic achievement and lifelong success. It followed that occupational class differences in the United States must reflect qualitative differences in the cultural-intellectual development of individuals that resulted from environmental factors as well as the influence of different genetic endowments. He argued that it was simply logical that at least some measure of the abilities demonstrated by the professional classes derived from heredity.[109] Second, he noted, whatever the precise source of differences reflected in the class structure, a well-established negative relationship also existed between fertility rates and cultural level. This correlation was itself a cause for concern. It meant that the rate of natural increase in groups with lower standards of living was higher than in groups with "superior" cultural levels.[110] Because economic class and culture-intelligence were positively related, while class and fertility were negatively related, class-based birth differentials were "probably the most injurious of all at present . . . retarding cultural advance throughout the whole country."[111] For each generation, because the lower socioeconomic levels contributed "more than" their "share to the next generations," it was likely that the proportion of population drawn from the superior cultural level would decrease.[112]

The continuation of this differential fertility pattern would have significant consequences for the social heritage of the nation, and likely its genetic endowment as well, "undermining our most precious inheritance, the capacity for high intelligence."[113] Moreover, "efforts to improve the environment [would] be retarded, and perhaps ultimately entirely offset, if at any time there [were] too many individuals in the population incapable of responding to an improved environment." In the United States of the thirties, Osborn noted, "among groups at different socio-economic levels . . . education and social improvement are handicapped by the high proportion of children at the lower levels, so that efforts at improvement are retarded with each new generation."[114] Further, "a vicious circle" became "apparent" in the relationship of cultural level and fertility. Lorimer and Osborn noted that *"retarded development fosters the maintenance of excessive fertility; and excessive fertility tends to retard economic and cultural development."*[115] Thus, through their logic connecting class, culture, and family size, fertility becomes a naturalized measure of quality and social worth, and the logic of eugenic better breeding comes back into population control discourse through socially generated differential rates of so-called natural increase.

And, although Osborn's arguments located the dangers of differential fertility in cultural difference, these differences were still racialized. It "is not biological heredity but the insurmountablility of cultural differences" that sustained the logic of racial distinctiveness. Thus, deploying another device of cultural differentialist racial discourse, Osborn did "not postulate the superiority of certain groups or peoples in relation to others but 'only' the harmfulness of abolishing frontiers, the incompatibility of life-styles and traditions."[116] This logic underwrote his argument that cultural difference militated against racial inclusion. As he saw it, with regard to "more or less distinct races with separate cultural patterns," the "problem of cultural assimilation overshadows any eugenic aspects" of racial problems.[117] "Their distinctive physical characteristics make them difficult to assimilate, even if racial assimilation were desirable." But "serious cultural problems" emerge with the question of assimilation because the significant differences in "their present cultural qualities, standards of education and sanitation are such as to complicate and retard the development of adjoining white groups." Biological differences did not legitimate racial segregation and immigration restriction; cultural differences did. As Osborn concluded, "Policies which would restrict births in some large groups and encourage them in other large groups" could not be "defended on biological grounds"; the scientific evidence was lacking. However, such policies "might appear wise and necessary" if "a certain culture [would] be thereby strengthened" or if "certain groups provide[d] a superior environment for rearing children."[118] Thus the new cultural model of difference reinforced longstanding tropes of biological racism that justified "the same old acts" of social regulation.[119]

Population dynamics ensured that race differences would continue to be a "matter of grave social concern, since racial problems" were, in his view, "accentuated by the tendency of minority groups to increase at the expense of the majority."[120] Lorimer and Osborn concluded that "suspension of judgment" on the question of innate racial differences was "made easier" because there were "no great differentials in reproduction rates among large racial and national groups in the United States."[121] Therefore, whatever genetic differences did exist, their relative proportion within the population would not change much. Increased differentials would, they predicted, lead to greater social conflict.[122]

As the nations of the Global South gained political independence in the fifties, Osborn shifted his focus to a global framework of differential fertility in which the higher fertility rates of former colonies were of great concern. Yet even before World War II, he expressed concerns about global population trends. He concluded that "high birth rates [were] making for continuous over-population in large portions of the world" such that "world population trends," in his view, were a "continuous threat to the orderly advance of humanity."[123] These trends were a threat because "to have the most civilized part of each nation, and of the world, declining in numbers because of a low birth rate, while the least civilized part is still rapidly increasing because of a reduced death rate, is to produce dislocations which would be dangerous to civilization if they should continue for too long."[124] Moreover, he noted, "there seems little prospect for eugenic progress in countries where the great mass of the people are in a state of intense poverty and ignorance as a result of over-population and cultural retardation."[125] Thus, while Osborn discounted the significance of heredity in comparison to earlier eugenic claims, in his view, heredity accounted for a non-negligible, if still uncalibrated, proportion of difference that linked to and could be tracked by cultural differences in population dynamics.[126]

Osborn's writings about race relations engaged one final constitutive feature of cultural differentialist philosophies of race. As Balibar notes, taking up the arguments of cultural differences advanced by anti-racisms of the 1930s, cultural differentialist discourses double back to configure new grounds against racial inclusion by "naturaliz[ing] racist conduct" as cultural chauvinism.[127] Lorimer and Osborn situated racist reactions to difference in naturalistic emotionalism and intolerance. As an example, while Lorimer and Osborn pursued objective science and found no biological race differences, they noted that it was not "easy to obtain a result free from personal bias, for we are dealing here with topics which touch deeply the basic racial, social, and local prejudices of mankind."[128] This doubling back to deploy cultural chauvinism in affective economies of race is most baldly apparent in Osborn's discussion of the effect of differential fertility, which played on these emotional prejudices:[129]

> The eugenist might say . . . that it makes no difference whether the world is finally inhabited by people who speak English, Chinese, or German; whether the color of its people be white, yellow, black, or red; whether we have a culture of Christian democracy, or that of the Emperor-God. That all these things are less important than the development of human beings whose genetic potentialities for sound growth and happiness surpass those of any group of humans alive today. But public opinion generally would not acquiesce in such a position. We shall continue to care that our own culture, our own language, our own people would survive.[130]

Even as they declared that claims of genetic difference were unsupported by scientific evidence, Osborn's writings participated in and directly contributed to the shift to cultural differentialism that reinvigorated racial ideologies in the mid-twentieth century.[131] Osborn's discourse fed into and on cultural anxieties about national, racial, and class differences. Thus, his references to the prejudice and emotionalism of earlier eugenicists and those who supported biodeterminism served primarily as a rhetorical device by which his arguments appear more scientific and more moderate. But the configuration of race, class, culture, and fertility rates in Osborn's writings resonated strongly with earlier eugenic concerns, such as Karl Pearson's lamentations that a disproportionate fraction of each generation was born to lower classes.[132] As the eugenic narrative of biological difference receded, the Malthusian scale of modernity reemerged as the means by which reproductive performance would be used "in thinking the difference between cultures."[133]

Although Balibar notes that "*racism always presupposes sexism*," he does not employ gender as an analytic tool.[134] Alexandra Stern, however, illuminates specific mechanisms by which the racialized logic of eugenics—and I would add cultural differentialism—naturalized and racialized social gender differences. In particular, she concludes that US eugenics intertwined "the health and survival of individual, nation, and western civilization" through a "family-centric" narrative that "rested on and demanded sex/gender uniformity."[135] That is, the theory of culture taken up by eugenicists in the midcentury conceptualized the family as the fundamental ground of habit and attitude formation. The normative family socialized good citizens, while deviations caused social pathology (crime, poverty, juvenile delinquency at home, and population explosions abroad). As Osborn noted, "There are no influences on the development of the child that are more decisive, more deeply set, or more permanent than those derived from the home in which the child is reared." He continued, "Chief among all the factors of home to mold the character of the children in it are the emotional attitudes, ideals, ideas, and standards of the parents, the primary place being held by the mother. Her attitude toward the young child carries more effect than does any other single influence. . . . For she contributes to her child's mental, social, and physical development in thousands of different ways." Following Malthusian logic, he concluded, "Often indeed the child's permanent stability and character are made or marred in relation to his mother's interpretation of her role."[136]

Osborn's family-centric cultural narrative renaturalized difference in essentialized genders, which were required for families to function effectively. By the fifties, Stern demonstrates, eugenics research helped configure the scales of masculinity and femininity to meet the demands of heteronormative marriage and the nuclear family. This gendered nuclear family became the fulcrum of racialized nationalism in social science and public policy.[137] For his part, Osborn recommended differential education for girls to encourage them to choose interests and occupations that would be compatible with heteronormative family life.[138] For demographers, the naturalized gender relations of the demographic family unit, measured quantitatively through fertility and contraceptive-use rates, provided the obvious linchpin between individual reproductive performance and group dynamics. By focusing on man's social heritage, vested in the (racialized) heteronormative family, the cultural differentialism embraced by demographers would situate the discipline in the center of the hegemonic masculinity formation of mid-twentieth-century US nationalism.[139] With Osborn's backing, this neo-Malthusian masculinity traveled the world.

Osborn's agenda for both domestic and international population policy derived from the gendered logic of his cultural differentialist eugenics. The earlier generation of eugenicists generally opposed birth control. Osborn disdained affiliation with the birth control movement, but he supported the dissemination of contraception by "competent" agencies.[140] In his view, the strong negative relationship between cultural level, socioeconomic status, and fertility made voluntary fertility control a necessity. As he reiterated in *Preface,* low fertility was associated with "high economic levels and civilized achievement," and high fertility was associated with "ignorance and isolation." He did not think it was reasonable to expect the successful classes to give up birth control, because birth control itself was "only a method." The increase in its use derived from "the culture pattern of the small family," which was a product of "contemporary life." That is, the small-family pattern emerged in "the modern world" as death was brought "relatively under control" and child rearing became more costly and time-consuming.[141] He continued, "Now that deaths were relatively under control, the distribution of births was the important factor . . . births will determine what races will survive, what language groups will increase, which regions of the world will feel a pressure of population and which will be under-populated, which cultural groups will grow and which will decline." In his analysis, in the mid-twentieth-century United States, the small-family cultural pattern was found primarily among the superior groups. Therefore, in each generation in which other groups remained unassimilated to the small-family pattern, the proportion of individuals who possessed superior qualities would likely decrease. If this situation continued, he stated, differential fertility might "seriously retard cultural advance." It was wiser policy, therefore, to promote contraceptive use among the less successful classes, races, and cultures than to try to suppress its use. "Instruction in the use of birth control," he argued, "is thus a necessary part of other efforts to improve social conditions." It should be "absorbed into the accepted social habits of the country." He continued, "The change from uncontrolled reproduction to conscious control of births is essential to a rational world civilization."[142] It was an essential cultural competency of the modern age.[143]

Because the correlation between occupational status and intelligence was also strong, specific measures should be taken to direct public opinion toward the voluntary limitation of family size among those who were economically and socially ill equipped to care for large families. He hoped that mass education in the biology of heredity, the small-family norm, and birth control techniques would shape public opinion and encourage people to have the number of children appropriate to their genetic and environmental situation. With regard to any racial differentials, he argued that "an acceptable eugenic program . . . would tend to equalize any disproportions . . . between the natural increase of whites, blacks, Indians and Mexicans." Incentive programs to encourage the "right people" to increase their birth rates would be necessary, a cause Osborn championed personally.[144] In particular, he was concerned that the crowded conditions of urban areas would discourage higher fertility among the "right people." In addition, the goal of his gendered educational proposals was to help women adjust to their proper roles and promote "birth release," the higher fertility that he saw as an obligation of the more fortunate, educated, and civilized.[145] In the case of rapid population increase in the Global South after World War II, the initial solution recommended by Osborn would promote fertility control values to assimilate the populations of emerging nations to the small-family cultural pattern as quickly as possible.

The Demographic Establishment

Osborn's close working relationship with Frank Notestein and Frank Lorimer positioned him well to promote population science.[146] Together they built three of the most influential sites of midcentury demography: the Office of Population Research (OPR), which guided the development of demographic theory and method; the International Union for the Scientific Study of Population (IUSSP), demography's international professional association and cosponsor of UN population conferences;[147] and the Population Council, which dominated international demographic research funding and training through the mid-1970s. In these sites, they captured authority for fertility control and dispersed midcentury demographic knowledge and knowledge-building techniques globally, effectively collaborating in creating the world in which the numbers worked.[148]

Throughout the period of its ascendancy, demography remained a gentlemen's club. Osborn and his associates continued to police the boundaries between their science and family planning organizations, "jealously guard[ing] its autonomy."[149] To avoid being associated with birth control, Frank Lorimer withdrew from the effort to reorganize population experts after the war when it coalesced into the International Planned Parenthood Federation (IPPF) in 1952.[150] At the same time, the initial invitation list to the Population Council's founding meeting included only white men of the Global North. As the date approached, two white Western women were added to the list, but no woman served on the council board for many years.[151] The founding meeting, which took place as the IPPF was being established, included the now routine disparagement of birth control advocates. The sentiments expressed were that "the 'radical birth controller,' while fulfilling a useful task, lacked a proper understanding of the necessary policies for particular countries at particular stages of economic and social development."[152] The council was organized so that its members would be "selected according to public esteem, professional competence, and interest and willingness to promote relevant investigations."[153] The distancing gestures of demographers reflected their belief that action was still premature. More research was needed to determine which actions should be taken. They continued to prefer a more quiet approach, working "under the banner of study rather than persuasion" to move the debate.[154]

As the Population Council's director, Osborn incessantly emphasized that the council was not an action organization. He routinely declined invitations that might publicly associate him with Sanger or birth control organizations, always reiterating the difference between science and action. As he explained on one occasion, "It is simply that having assigned to the Population Council the scientific rather than the action field, we feel it is wiser not to appear very much in public on the action side, even though some of the things we are working on are pretty revolutionary."[155]

Although Osborn and his compatriots maintained public distance from the birth control movement, they nonetheless engaged with birth control organizations behind the scenes. In particular, when they had the opportunity, Osborn and Notestein endeavored to steer the direction of the PPFA and the IPPF, shape their access to resources, and influence their leadership selections. In the 1930s, Osborn participated in the Council on Population Policy, which he convened in an effort to manage relations between various lay population organizations. He also participated in recurring discussions about the possibility of a merger between the American Birth Control League and Sanger's organization; the 1942 merger resulted in the Planned Parenthood Federation of America (PPFA).[156] Throughout the 1950s, Osborn sought to influence the leadership of the PPFA toward a less aggressive approach. In setting up his policy discussion group at the Population Council, the Ad Hoc Committee on Policy, Osborn included PPFA participants "*without* fixed positions."[157] As the IPPF coalesced, he privately expressed reservations about Sanger's leadership, generally linking the leadership question to the need for a calmer, more scientific approach to the field.[158] At the end of the decade, Notestein tried to persuade Alan Guttmacher to work for the Population Council instead of taking a leadership position in the PPFA.[159]

The most sustained engagement of Osborn and his associates with family planners developed between the medical divisions of the PPFA and the Population Council. Built from the base established in the 1930s, when birth control clinics tested the effectiveness of various contraceptives, joint projects and conferences laid the groundwork for the contraceptive quid pro quo between reproductive scientists, family planners, and demographers.[160] Although he would not publicize them, Osborn encouraged these connections practically. When declining requests for the Population Council to become involved in applied research, he often referred inquirers to Planned Parenthood's medical division.[161] Through small grants, a discreet word to funders, and quietly cosponsored luncheons and seminars, Osborn facilitated the expansion of PPFA contraceptive product testing, eventually making the PPFA medical division a site of applied research that tested contraceptives with the support and encouragement of the Population Council.[162]

Throughout this period, Sanger's continued presence on the public stage and involvement with population issues was a particular source of irritation, and the Population Council leadership tried to manage her influence and impact. For example, in 1948, citing the possibility of controversy and the normal process of population change through modernization, Frank Notestein advised the US military not to allow Sanger to visit Japan.[163] In another case, Osborn managed the possibility of controversy by undermining Sanger's plans for an IPPF conference in Washington in 1958. In letters to IPPF headquarters, he relied on the Population Council's reputation for discretion and serious purpose to persuade IPPF officials in Britain that the conference would only stir up religious and political opponents, and would fail to secure increased funding for the IPPF, which was a central goal.[164] In addition, although it quietly provided financial support to the American Eugenics Society, the Population Council declined to fund the IPPF during the years Sanger and Dhanvanthi Rama Rau led it.[165] Since the council was a strong voice on financial resources for population control, its reluctance to provide funds to the IPPF sent a negative signal to other potential funders and had a negative impact on the organization's efforts to gain a solid financial backing.[166]

This was not simply about Sanger; rather, the Population Council leadership was engaged in a gendered power struggle for authority within the reproductive arena. This struggle shaped population scientists' public distance from and private interventions into birth control organizations at their margins. When Sanger decided to step down as president of the IPPF, gender politics were again at play in the selection of a successor. Noting that "it would be a great pity not to go along with Mrs. Sanger's desire to resign," Osborn urged his old friend, the British eugenicist Carlos Paton Blacker, to take on the presidency. In making his case, Osborn argued that the "selection of a president would affect grant-giving bodies," and he felt a "higher grade head man" would be best. "If the IPPF can't find just the right man and substantial funds now, it might be a good thing to give it wise and less aggressive direction for the next two or three years . . ." As president, Blacker could "develop financial possibilities" and "be looking around for a younger man to succeed" him. Osborn noted that this strategy had been effective at the Population Council, which was poised to "move on into a much larger operation."[167]

Throughout the several letters between Osborn and Blacker about the matter, both men "naturally" assumed the next president would and should be a man, especially if the IPPF hoped to increase its influence on the world stage. They were not alone in these sentiments; Hugh Moore also expressed more confidence in the organization as Sanger stepped down, noting that it had come to "seem more than a woman's organization."[168] Sanger also conceded that "if we elect a presentable man, rather than a woman, he will be more likely to influence and bring into our movement" prominent men active in world affairs.[169] Although a woman was elected as the next president, the IPPF in the 1960s would be more than a woman's organization. Blacker and a succession of prominent men would serve in executive positions within the organization.

Beryl Suitters, who chronicled the early years of the IPPF, noted that in the 1960s "family planning" was "taken seriously, rather than dismissed as a quackish notion or an obsession of interfering feminists." She commended the change, noting that the IPPF had "to lose some of the feminism" because "in many areas the men were the ones to initiate and sanction change and masculine goodwill was vital to any effective work." Suitters was referring to the sexism of men of "other" cultures, but it is an apt description of the position of family planning organizations in relation to Western masculinities.[170] The issue was never just that family planners were on the action side of the science/action boundary; it was always also that their actions were misguided by feminism. As Ellen Chesler has noted, by this point, population scientists were "baffled" by Sanger's continued faith in women, seeing it as a remnant of "kooky, sentimental feminism." This view extended to birth control movement organizations as well, in which, demographers suggested, the "'feminist bias' was subverting family planning programs in the developing world" because their approach, which emphasized the improvement of individual maternal and child health through clinic services, was "too expensive" and "likely to fail."[171] They did not believe as Sanger and family planners did that ensuring women's access to reliable birth control would be sufficient to move aggregate trends in the right direction.

As Osborn suggested in his letter to Blacker, the Population Council was ripe to become a much bigger operation. In the 1960s, it would play an increasingly prominent role in the development of population control programs. The guiding philosophy of population control—grounded in demographic facts and Population Council funding—would realign individual women's need for and access to contraceptives to meet aggregate goals. Osborn would attend the 1959 IPPF conference in New Delhi, where Sanger finally retired. His participation at the conference was symbolic of an important shift. By that time, he believed that "informed opinion" was catching up "with the need for action."[172] In the 1960s, as US money and perspectives played an increasingly larger role and the organization's influence grew, the population control perspective came to dominate the IPPF agenda. This shift was never complete or without controversy, but it was dominant, nonetheless.[173]

• • •

Over the course of the 1930s and 1940s, the "amorphous field of population" was configured into the social worlds of mid-twentieth-century demography through contests that left birth control advocates outside, at the margins of the discipline, and drew eugenicists into the heart of the discipline. The explicit exclusion of Margaret Sanger and the "cause" of birth control set the initial boundary of the field. Denial of its connections with birth control was used successfully in establishing the scientific status of demography before statistical tools proved their worth. Those denials hinged on masculinist scientism to represent demographers as calm, moderate, and rational men whose authority derived from careful study and birth controllers as emotional propagandists who needed to be constrained by the demographic facts. This characterization of demographic science and birth control advocacy aligned with conventional affective economies of normative gender differences to amplify the authority of demographers. Demographers' ritual retelling of the foundation story reconfirmed the credibility of their science and diminished that of birth control advocates. In this sense, the founding tale captured the "institutionalization in motion" by which the field of demography and population control were configured and bounded by the Sanger problem.[174] The conventional wisdom, inscribed by demographic narratives, is that Sanger had to be isolated because the publicity her propaganda generated would harm the goal of developing reliable knowledge with which to manage aggregate fertility trends in the public interest.

In most accounts of the founding meeting of the PAA, Sanger is portrayed as a silent, passive figure. Marginalized in the founding meeting, indeed, she did not become a member. She did not support the PAA financially, nor did she lend her name to it. These actions, among others, displayed an unwillingness to defer her agenda to that of population scientists. Moreover, Sanger was not blind to the gender politics at work in this instance.[175] This episode, one of many in which she was used and sidelined by prominent men, contributed to her resentment and deep ambivalence about the male leadership in the birth control and family planning movements.[176] Margaret Sanger was certainly a complex person of strong opinions and personality, and she has been judged harshly in the historiography of birth control. However, the characterization of her as an attention-seeking troublemaker rests on the assumption that withholding recognition of her professionalism and expertise was justified; she was after all a laywoman.[177] Rather than simply accepting this characterization at face value, interrogating the claim and the stories by which it is deemed credible illuminates how and for whom the stories work.[178]

In this case, as the chapter has shown, the boundary work by which demographers excluded birth controllers from authoritative positions within demography's professional organizations and research centers also enabled them to legitimate their scientific management of family planning in the interests of population control.[179] The credibility contest left the birth control movement on the wrong side of the division between social science and political advocacy. Without the imprimatur of science, the movement and its feminist remnants continued to be cast as a special interest group, driven by ideology, and thereby its individualist claims were discounted in relation to scientific claims about aggregate population effects. Movement organizations struggled throughout the midcentury to achieve recognition as legitimate public health agencies. By the time birth control organizations did earn legitimacy in the 1960s, they were securely in the hands of a demographic science authorized to respond to the aggregate crisis of rapid population growth.

Viewed in this light, the figure of "Mrs. Sanger" in the founding story is a "sticky" boundary object against which demographic credentials were displayed. The contrast with unruly birth control advocates was indispensable; it strengthened demographers' claim to be the credible arbiters of fertility control. The figuration of Sanger and family planners as excessive, even dangerous, deflected and deflects attention from the interventionist agenda of demographers. Frederick Osborn, the hero of the founding tale, used the calm quiet voice of moderation to weave the perspectives and goals of social eugenics into the epistemological and organizational fabric of the discipline. He helped the discipline capture controlled fertility for a global hegemonic masculinity that naturalized cultural difference, ranking men and nations based on their fertility rates. Under his leadership, the focus of better breeding shifted from the quality of the genetic contribution of large racial groups to the social determinants and quantitative impacts of rapidly growing cultural groups.[180] In this frame, those declared to be superior were those whose rate of reproduction adhered to neo-Malthusian principles of the small-family system. In the light of the "frightening statistics of population growth" produced by demographers in the 1950s, the defense of civilization against irresponsible increases in numbers would justify the use of contraceptive technologies for population control.[181]

Cast as emotionally driven, the troublesome women of birth control and family planning organizations have also been made to bear the weight of the prejudices that underwrote population control programs. There is no question that the trade-offs made by birth control advocates in the pursuit of the goal of securing women's access to birth control resulted in deeply problematic practices that continue to haunt reproductive justice activism. But the enduring representation of Sanger and family planners as the principal source of the racialized politics of population control owes a great deal to the credibility contests that configured the population control establishment of the mid-twentieth century. Sanger and family planning advocates did become an eager and sometimes injudicious audience for demographic statistics and policy recommendations. They, like others in the downstream publics, were affected by the panic of numbers demographers produced. Full accounting of the injustices wrought by international family planning agencies is critical. But to conclude that they might have prevented the dominance of the population control perspective, in my view, underestimates the force of the cultural and epistemic authority of population science. The strength of the population control argument is perhaps indicative of the gendered disadvantages family planners faced in the contest for credibility all along.[182] In part, demographers were more successful in sustaining their claims than were family planners because the space they configured was "better able to hold the diverse interests of the powers-that-be."[183] For birth control organizations, disparaging caricatures of planned parenthood nuts and interfering feminists simply hastened the declining influence of women's rights arguments in public debates at midcentury. Thus while demography helped to legitimate contraceptive technology, it did not extend that legitimacy to individual women's perspectives on fertility control.[184] Within the demographic narrative of population control, as one later feminist health activist would note, "it was as though women weren't human."[185]

The initial boundary work of demographers rested so heavily on the exclusion of women's advocacy that it conditioned the field's resistance to feminist movements and scholarship later in the century.[186] And the discipline's gender politics earned it a strong reputation for sexist attitudes and behavior among the few women in the field in the midcentury.[187] However, while epistemology and methodology always bear the ghosts of the historical context in which they are configured, they cannot be reduced simply to personalities and organizational structures. Intellectual work matters. As Notestein observed, research findings led the way. Within the bounds of the discipline, demographers produced knowledge and measurement tools that drew on and participated in the gendered coloniality of power that warranted the social science discourses of the mid-twentieth century. The next chapter investigates the positioning of women and other others in emerging demographic practice and epistemology by examining the early demographic research into the impact of contraception on fertility patterns. It elucidates the empirical grounds of a new masculine discourse of reproductivity and population control. This discourse displaced women's desires with "bloodless reified categories" of aggregate population figures and was "animated by wonder and vexation" at the rapidly expanding numbers.[188]

4

Remaking Malthusian Couplings for the Contraceptive Age

> The practices of thinking and writing that are of special concern . . . are those that convert what people experience directly in their everyday/everynight world into forms of knowledge in which people as subjects disappear and in which their perspectives on their own experience are transposed and subdued by the magisterial forms of objectifying discourse.
>
> —DOROTHY SMITH

IN THE UNITED STATES IN THE 1930S, POPULATION SCIENCES COA-lesced into a scientific community that marked its borders and warranted its knowledge through a narrative that loudly distinguished it from the woman-led birth control movement and quietly embraced eugenicists. Within the boundaries of the resulting gentlemen's club, research proposals and drafts of scholarly articles, charts, and graphs circulated for comment and revision among a small circle of scientists and benefactors. Through these circuits, demographers configured the standardized measurement practices and numeric representations that constituted "the facts" about past, present, and future population trends worldwide. These "facts" "coordinated the activities" of the discipline through the next several decades.[1] They also commanded the attention of the popular imagination about fertility and its control. The emerging conceptual practices of demographers drew on and participated in the sociological imagination that coalesced in the mid-twentieth century around structural functionalism's grand theory of social action and the quantitative methodologies of scientism. Demographers contributed two interconnected objects to the sociological imagination: a neo-Malthusian couple reconfigured for the contraceptive age and a neo-Malthusian statistical remapping of civilizations arrayed by the contraceptive (in)competency of their populations. Demographic figurations of pregnancy risk, natural fertility, and contraceptive competence of the cosmopolitan neo-Malthusian couple underwrote the nuclear family normalized in midcentury sociological narratives on sex roles, family life, and economic development.[2] Such couples anchored the "American dream" and the modernization paradigm of US postcolonialist foreign policy.

Feminist sociologists have sharply critiqued midcentury social theory for rationalizing patriarchal family structures as necessary for social stability, and thus closing down discursive space within which to articulate or analyze gender-based conflict.[3] Similarly, feminist demographers have criticized their field's reliance on the conventional gender assumptions embedded in modernization theory.[4] Those critiques primarily focus on midcentury grand social theory. This chapter focuses attention on the quantitative practices wedded to that grand theory. By excavating the "pre-history"[5] of elements brought together by the statistical measurements developed and deployed in early contraceptive research, it illuminates how the numbers worked to normalize fertility control within both the hegemonic masculine narrative of modern marriage and the modernization paradigm.

Throughout the 1920s, debate swirled among "experts" and "laypersons" about whether or not contraception actually worked. That is, did it prevent pregnancies that otherwise would have occurred? For the new population science, the primary concern with contraceptive effects was whether the observed decline in fertility rates resulted from increasing contraceptive practice or from "racial degeneration," defined as a decline in the underlying organic ability to reproduce. What little information existed about contraceptive effectiveness came from birth control clinics, which men of science disdained. Into this mix in the 1930s, Raymond Pearl and Frank Notestein contributed a number of important studies that provided statistical "proof" that contraception was indeed effective and likely contributed to the decline in aggregate fertility rates. These landmark analyses "essentially disposed of biological interpretations of fertility differentials and trends."[6] In so doing, they defined "what qualities or characteristics should be measured . . . [and] . . . what units of measurement should be used" to construct authoritative knowledge of both uncontrolled and controlled fertility, and produced (racialized and classed) models of pregnancy risk and competent contraceptive practice.[7] Therefore, they are an important site in the genealogy of demographic categorization that configured the modern population imaginary. The chapter excavates the intertwined differential and hierarchical gender logics underlying the demographic figurations of pregnancy risk, natural fertility, and controlled fertility that produced the feminine figures of (in)competent contraceptive practice arrayed by race, class, and cultural categories.

Because contestation with birth control advocates shaped the foundations of the discipline, it is useful to reread these foundational demographic studies in conjunction with the movement accounts that demographers disparaged. Doing so reconnects those demographic facts with the situation of their production.[8] Thus, the chapter begins with a fresh look at letters and testimonials published by birth control advocates as evidence of women's experiential knowledge of fertility control and its failings. Drawing on the work of Lauren Berlant, I read these sentimental narratives as written into and organized by the "intimate public" of women's culture.[9] As such they speak to the "fleshy world"[10] and "the structure of feeling"[11] in which contraceptives circulated and practices acquired meaning for women negotiating modern marriage.

Of course, in the view of demographers, the sentimentality of these texts is precisely what marked them as propagandist. As we saw in the last chapter, the charge of emotionalism was a critical tool with which Sanger and birth control were dismissed. However, reading demographic narratives of pregnancy risk through the lens of women's narratives illuminates the local gendered risks erased by the demographic understanding of fertility and its control. As a backdrop for the subsequent close reading of demographic studies, women's stories "unsettle the . . . lines that delimit the zone of knowledge" established by the demographic figurations.[12] Specifically, by reading the demographic texts through the lens of movement literature, both of which were based on listening to women, the subjugated knowledge contained in women's narratives but missing from demographic figurations comes into sharp relief. Attention is thus drawn to the aspects of "women's perspectives on their own experience" and engagements with ideologies of marital love, sexuality, and procreation that are "transposed and subdued" in the bloodless sociological facts of contraceptive need.[13] These demographic figurations, I show, suppress gender-specific disjunctures between the ideologies of romantic love, companionate marriage, and the actualities of women's experiences of marital intimacy and prudential procreation. After World War II, the demographic configuration of pregnancy risk and contraceptive competence built from demographic figurations of metropolitan women's contraceptive practice would "cover the entire surface" of the world "that stretched from the body to the population."[14]

Finally, although early contraceptive-use studies reaffirmed that women wanted and needed access to contraceptives, they also performed some additional boundary work by deriding the women-led birth control movement. In his account of controlled fertility, Frank Notestein argued that while contraception had reduced fertility, the movement's activities had played an insignificant role in that trend. Moreover, he concluded that practical access to contraceptives was not as important to contraceptive effectiveness as the motivation to acquire and use them. By the end of the decade, demographers had succeeded in securing the scientific understanding of human fertility trends to the priorities of the discipline. Within this regime of knowledge, birth control clinics became sites of applied research, subordinate to and separate from objective science and policy institutions.

Women's Narratives of Contraceptive Need

In 1928, Sanger published *Motherhood in Bondage,* a compilation of nearly 500 letters sent to her requesting contraception.[15] In 1930, Lella Florence published *Birth Control on Trial*, an account of her work with the birth control clinic in Cambridge, England. Both provide valuable insight into the experiential knowledge of "the fleshy world" that shaped women's engagement with birth control. That is, if one thinks of pregnancy as the result of "worldly practices" enacted in intimate spaces in which bodies, institutions, ideologies, and technologies are co-constituted, then the early twentieth century in America was a period of sustained challenge to previously settled reproductive arrangements.[16] Institutional supports for the family shifted as people moved away from their place of birth and entered new occupations. Women also tested the confines of conventional femininity by participating in public life through social and political movements. At the same time, dominant masculinity underwent changes that undercut the regime of Malthusian prudential restraint. The competing figures of nineteenth-century American masculinity—the civilized Christian gentleman who forswore personal pleasures and the aggressive, vigorous man of sporting culture who heartily indulged all his appetites—came into increasing conflict as masculine sporting culture expanded with the rise of consumerism. Moreover, despite the always-greater restrictions on them, women began to participate in these "cheap amusements" as well.[17] In this context, the conventional Malthusian discourse of prudential restraint became increasingly unworkable.

Sigmund Freud, the father of modern theories of sexual desire, based his theories of sexual repression in part on his observations of patients' frustrations with the demands of prudential restraint. His personal journals also recorded the struggles within his own marriage to reconcile sexual restraint and intimacy.[18] In the early twentieth century, male sex radicals such as Floyd Dell and Havelock Ellis used Freud's theories of repression to oppose Victorian (hetero)sexual restrictions.[19] Birth control and free love advocates such as Emma Goldman and Margaret Sanger likewise spoke against Victorian standards of (hetero)sexual restraint. They discussed marital sexuality in terms of romantic love as they argued for women's greater sexual rights, emphasizing that women had the right to sexual expression free from the fear of pregnancy.[20]

From the perspective of hegemonic masculinity, however, the prospect of women's sexual independence was a problem; it threatened to exceed public (patriarchal) institutions meant to secure men's confidence in their paternity and their implicit right of access to women's bodies.[21] The marriage reform movement of the 1920s redomesticated the more radical critiques of Victorian morality. It promoted a model of "companionate marriage" built on romantic love and sexual communion to preserve and improve marriage as the basis of modern family life and citizenship. Marital advice literature of the 1920s through the 1940s "attempted to meld equality for women with sexual intimacy as the essential cement of modern marriage," arguing that sexual intimacy strengthened the psychological bond between spouses. However, the discourse of companionate marriage also renewed and preserved male-dominant heteronormativity in that husbands retained the power to initiate sexual relations.[22] It resolved the competing values of restraint and vigor in modern middle-class masculinity in favor of sporting culture. The greater sexual expressiveness in modern marriage demonstrated the vitality of modern men, thus defining masculine libidinal patterns as the norm. In what Christina Simmons calls the "flapper marriage," the marital advice literature promoted egalitarian marriages between "vital, modern, yet pliant flapper sweethearts . . . [and] sensitive, yet still masterful men." America's sweethearts, modern girls who were chaste before marriage, were "good sports" after marriage. Willing "to put [their] man first," they would become active sexual partners with their husbands' guidance. Modern husbands were to be tender, considerate, and help with the housework. But men were still in charge, especially in the bedroom.[23]

Although the discourse of companionate marriage undermined the Malthusian doctrine of prudential restraint, it fully accepted the Malthusian economic logic: one still needed to limit one's family to a size that one had the means to support. Such restraint demonstrated not only one's class standing but also one's competence as a citizen. Excessive fertility was still seen through a Malthusian lens as a drain on society, producing poverty and dependency. Thus the heteronormative promise of "companionate marriage" and "modern family life" hinged on effective contraceptives to reconcile the competing demands for more sex but fewer children. Effective birth control, however, was still elusive, and women were left to negotiate the gap between the promise of marital intimacy and the lived reality of uncertain birth control.[24]

Organized by broad themes, the letters compiled in *Motherhood in Bondage* expressed women's desire/need/desperation for *something certain* to prevent pregnancy. Sanger framed these "human documents" as powerful evidence of the unjust and often dire consequences produced by legal restrictions on access to contraception.[25] The title's allusion to slavery and the reading posture called for by the introduction announce the book as a text of the sentimental genre of "women's culture." As Lauren Berlant argues, the "intimate public" of femininity organized an imagined belonging and directed it toward the evocation of sympathy for the misfortunes and injustices of women's lives, even as it maintained fidelity with conventional femininity. According to Berlant, sentimentalism is characterized by "the centrality of affective intensity and emotional bargaining amid structural inequality, and the elaboration and management of ambivalent attachments to the world as such." In particular, the texts of women's culture "foreground witnessing and explaining women's disappointment in the tenuous relation of romantic fantasy and lived intimacy" that "marks the scene of the reproduction of life."[26]

Sanger used the conventions of sentimentalism to position the letters as evidence of all women's suffering and demanded political intervention to end it.[27] As she noted: "If the great, prosperous and generous American public is brought to a realization of the tragedies concealed in the very heart of our social system, I am confident that it will recognize the importance of measures advanced which aim to prevent their recurrence in the future."[28] With this statement, Sanger sought to rouse readers to action by "harness[ing] the power of emotion to change what is structural in the world."[29] This goal was explicit in the comparison Sanger made between the power of the personal stories of suffering and "anemically intellectual" "academic" arguments: "An easy and even a pleasant task is it to reduce human problems to numerical figures in black and white on charts and graphs, an infinitely difficult one is it to suggest concrete solutions, or to extend true charity in individual lives. Yet life can only be lived in the individual; almost invariably the individual refuses to conform to the theories and the classifications of the statistician."[30] The "essential value" of the compilation of women's letters, she continued, "consists precisely in the uncanny power of these naïve confessions—unguarded, laconic and illiterate as many of them are—to make us see life as it is actually lived close to the earth, without respect for the polite assumptions and conventions of sophisticated society."[31] She concluded, "This power to bring the reader close to the heart of motherhood in bondage makes these records dramatic." Moreover, she argued, "they are as poignant in what they leave unsaid—but which we can read between the lines—as in their amazing revelation of fact."[32] Finally, she asserted, "these documents give us something that we cannot obtain from any number of 'maternity surveys' or 'biometric' computations of vital statistics—the secret, never-told factor of maternal anguish and sacrifice."[33]

As Lauren Berlant and others have shown, the genre of sentimentalism is fundamentally structured by the priorities and prejudices of the white bourgeoisie who produced and consumed it.[34] This cosmopolitan paternalism is clear in Sanger's framing of the "tragedy" of "slave maternity" above, and in her characterization that it was greatest among the urban poor and "Southern farm" women.[35] It is also apparent in the thematic categorizations of the chapters, which include, among others that reference health issues, "Girl Mothers," "The Pinch of Poverty," "The Struggle of the Unfit," and "The Sins of the Father."[36] Sanger's framing of the letters within the hierarchical conventions of gender-marked sentimental discourse should alert us to be attuned to the effects of those conventions in the selection and presentation of the letters. For instance, Sanger notes in the preface that she excluded letters that contained simple requests for information. The included letters resonate with Sanger's call to readers, explaining the writer's situation in sentimental prose. However, the letters themselves should not be dismissed because of the book's framing of them. They are powerful evidence precisely because they evoke strong emotions. They offer an invaluable window into the fleshy world and structure of feeling that gave meaning and value to contraception for women who sought it out.[37] As one woman noted, "It feels good to have some one to tell this to for me."[38] Moreover, the life stories the letters convey exceed Sanger's editorial categorizations. They constitute a richly complex archive of US women's engagement with birth control as a means to negotiate the disjunctures between the lived experiences of procreative bodies and the competing ideologies, institutions, and technologies of modern marriage that aligned them. As such, these texts serve to interrogate the disembodiment and dispassion accomplished through demographic measurement practices.[39]

Women often began their letters with a stark résumé of the "facts" of their reproductive lives—a statement of their age and an inventory of pregnancies, births, miscarriages, stillbirths, abortions, and infant/child deaths. Such inventories of the losses and near losses of their own lives and those of their "dear ones" were "proof" of their suffering and their need for birth control. As one woman noted, "I am thirty-five. In seventeen years of married life have brought eight children into the world and went down in the grave after three I failed to get."[40] Following the austere recitation of the facts, women offered poignant meditations on their efforts to manage the demands of conventional femininity. The letters portray women as worn down by the effort of balancing competing desires to maintain intimacy with and faithfulness of husbands, to "do right" by the children already born, and to secure health and longevity in their own lives. As the writer above continued: "My baby is nine months old and the thoughts of another almost kills me. Oh! tell me how to keep from having another. Don't open the door of heaven to me and then shut it in my face. Oh! please tell me, I feel like it's more important to raise what I have than to bring more."[41]

In seeking to "do right" by those already born, women often specified not only children's basic needs for sufficient food, clothing, and shelter but also their desire to give children an education as well as freedom from early wage work. As one woman noted: "I want to give my children a fair chance in this world. I have suffered from poverty all my life. My father died when I was two years old leaving my mother with four small children and penniless. I know what it means to go out in the world and earn a living without an education. Why should I raise a lot of children to have them suffer?"[42] As another concluded, "If no more babies come and time goes on there is a bright future in store for us."[43] Most poignantly, women expressed the desire to live long enough to raise the children already born, and the fear that another pregnancy would prevent it. "Two years ago I had dropsy when my last baby was born and the doctor told me I had to look out so I would not have any more as it would kill me, but the doctor was not allowed to tell me how to keep out of it, so you see I live in constant fear, so please have pity on me and tell me what to do as I have to live for the children's sake."[44] Another wrote, "My husband asked me to write you this letter since he realizes it is the only safe way to keep me for my little children's sakes."[45] Still another woman wrote, "There isn't anyone that could appreciate help or advice in any way more than I. I have five children, the oldest is seven years old and if I could just manage to not have any more I might live to see them or help send them through school."[46]

The specter of death in childbirth pulses through *Motherhood.* It is quite telling that women referred to the labor of childbirth as going "down to the valley of death."[47] Many of the women wrote as they anticipated an upcoming "confinement," as the period surrounding childbirth was commonly called.[48] It was not something these women looked forward to, especially as the number of pregnancies increased. As one letter writer said, "Now expecting the birth of a child any time. What on earth am I going to do? It is death to birth a child, let alone the other things after the child comes."[49] The risks of childbirth were great in this period.[50] Women expressed considerable anguish over the uncertainty of it all, not knowing whether they or the newborn infant would survive or thrive. Letters from the daughters and some sons and husbands who were left to raise their younger siblings after their mother's death in childbirth punctuate the risks women faced and the losses they suffered.[51]

An equally important theme that reverberated across the letters is the strain the risk of pregnancy imposed on marital intimacy and happiness.[52] The letters speak of the tender feelings between husband and wife suffocated by the demands of prudential restraint and/or trampled under the weight of "too many babies too close together." One woman noted, "My husband is numbered among one of the best and had done all he could for me, except we are both 'ignorant.'"[53] Describing marital tension, women often said that they "try to keep away from" their husbands, a colloquialism for the strategy of ensuring abstinence by avoiding affection and intimacy.[54] Women expressed that such strategies were bound to fail. As one letter writer explained, "I avoid my husband whenever I can, but any married woman understands a man." Another noted, "It is hard to hold a man's love and try to avoid close relationship." Such a strategy caused only greater strain and anxiety, often leading to quarrels or threats to "go elsewhere." As another woman noted, "I am married two years and have had two children sixteen months apart and it pretty near drives me frantic to always be worrying of more. I try and keep away from my husband but most always leads us to a quarrel and I am not strong."[55] Even when husbands acquiesced to abstinence, women worried. They worried about how long their own or their husband's resolve could last. As another noted, "I left high school at the age of seventeen to marry a poor man and I never have regretted it. I have done all my own work and borne my own children happily and with never a complaint, but I live in constant dread of another baby soon, and so does my husband. He has kept away from me for long periods but I cannot ask that of him forever."[56]

Another wrote after a self-induced abortion from which she nearly died, "Since then I have refused to have anything to do with my husband. He says he does not blame, but I don't know how long it can last. I know he can leave."[57] Another noted the cost to intimacy, writing, "My husband is trying to rise and I know I am a hindrance to him. I am cold to him because I fear another child if I warm up to him and even love him.

He loves me now I know but he says it is hard to be loving to an icebox and I know we are growing apart and it makes my heart bleed. I fear for our home."[58]

Many women also noted that self-imposed abstinence was difficult because they loved their husbands and desired intimacy. One noted, "I try to keep away from my husband but that's not right. I love him very much and we are young, but I have a constant dread all the time for fear I will get pregnant." Another wrote, "I am so afraid of becoming pregnant," but "what shall I do? I love my husband and we enjoy the sexual union. If only we knew of some good reliable contraceptives." One woman concluded, "I know it is impossible if a couple lives together and loves each other." Another noted, "My husband is so good and kind to me and would do most anything to keep me from becoming pregnant again but he is human and so good to me I like to do the best I can for him."[59] Some women, on the other hand, complained about selfish husbands who did not and would not restrain themselves. As one wrote, "But my husband has a head of his own and won't listen till it is too late, and then he gets mad tells me it's no fault but mine." Another wrote, "It is one thing certain, my husband won't give up his right as a husband, for I've pleaded for it as my very life seemed to hang on it."[60] In either case, good husbands or bad, marital affections suffered under the strain of trying to balance marital intimacy and procreative restraint.

Drawing on her experiences working in the Cambridge birth control clinic, Lella Florence summed up the impact of prudential restraint on marital affection and intimacy as follows:

> In the case of the older couples, work-weary and worn out, abstinence is apparently not a severe effort. But some of the younger women have told me what a strain it imposed upon both themselves and their husbands. The wife always felt the necessity of restraining any demonstrations of affection for her husband, and of meeting his affectionate overtures with coldness and rebuff, until gradually there grew up an icy barrier between them which both recognized, but which they could not alter so long as abstinence was enforced. If only these couples could have known of some certain contraceptive, they might have been spared much pain and much lost happiness.[61]

By *something certain*, women meant a means under their control that allowed easy intimacy, without the anxiety of awkward compromises. In particular, women sought an alternative to interrupted intercourse. That is, women often noted that their husbands were "good to them" and were "careful"—both references to withdrawal—but that this method was unreliable. As one woman opined, "I keep away from my husband and he is very careful with me but still it does no good." Moreover, such closely timed restraint was difficult, often making husbands "nervous wreck[s]."[62] As Florence noted, withdrawal was especially disruptive of tender feelings. "The whole act, which ought to be a happy extension of their love for each other, becomes a strained and miserable business, more often than not resulting in quarrels, ill-temper and worry, [because] the husband is under the necessity of being on the alert . . . and the wife suffers great anxiety lest he shall fail to do so."[63]

Women's requests for *something certain,* something a woman could count on to work, often also referred to the dangers and failure of commercial methods they had happened upon.[64] Speaking of the dangers of commercial products, one woman noted, "It is a constant worry for fear I get that way again and another horror of using any advertised appliances for fear of injury to one or the other of us." Another noted the unreliability of methods for sale: "I bought something from a lady but that done me no good. I got that way just the same. So another lady told me about a thing for men. But that failed and I got that way again." Another wrote, "I have tried dozens of preventives my neighbors advised only to find myself pregnant again." The cost of such failure was evident: "I thought I knew a safe method of Birth Control but four months ago for some cruel reason I found myself pregnant again with my fifth child. When I tell you my home is ruined I put it mildly."[65]

Finally, the phrase *something certain* also meant something trustworthy; doubt about the effectiveness of their efforts to prevent pregnancy was its own significant source of anxiety. Women wrote that waiting for proof was distressing in itself. As one letter writer noted, "I have lived in a state of suspense from one month to the next." Another observed that "the constant strain of not feeling certain of just where we are is very nerve-wearing on me and does not make home any cheerier and better for my family."[66] Florence noted that women often abandoned the diaphragm because they did not have confidence that it was placed correctly and stayed put.[67] A method that they could trust to work would relieve their constant fears and reduce the anxiety associated with receiving/responding to husbandly attention and affection.

The worry and doubt the letters convey signified an important disjuncture between the ideology of companionate marriage and lived intimacy.[68] Women blamed their lack of knowledge and/or access to reliable methods for the disappointing effects of their efforts to reconcile the ideals of romantic love with life-as-it-is-lived. Some letters also framed the matter as a desire for justice. As one woman wrote,

> So there is my story and here is my appeal to you. I love my husband dearly and he is very good to me but I am almost afraid of him to come near me and I can see that my fear of him is resulting in his paying less attention to me, and I don't blame him in the least because I can't be a real wife to him. I love him and my heart aches to be a real companion to him. He tries to make me think that it makes no difference to him but I know it does. . . . Is there no way for me to get this vital information? Must I just live—nay exist—in dread because an unjust law says I must?[69]

The letters expressed women's desires to "breathe and thrive," not just for themselves, but also for their husbands and children.[70] As one woman wrote: "If I could get one good contraceptive that would not fail but would be sure so I would not become pregnant till I can get strong again I would surely be glad. Then I could give my husband my true love and do my children justice."[71] And another noted: "In the hospital where my babies were born every woman was trying to find out the same thing. They asked doctors, nurses and each other. They were all in constant dread of more children. Such a condition is deplorable in this age of freedom in everything else."[72] As with sentimental literature in general, which affirms feminine conventions even as it contests them, letter writers desired something better within the domain of conventional femininity. As Berlant has noted, "In a sentimental worldview, people's 'interests' are less in changing the world than in not being defeated by it."[73] So too the women letter writers. They were not seeking an escape from feminine responsibilities. The promise of birth control was that it would enable them to balance competing demands of sexual intimacy and effective motherhood, marital and economic security.

Of course, the sentimentalism employed in birth control literature is precisely what population scientists and other critics disparaged. Because it appealed to the heart, it could not be counted as objective. This sentimentality rendered birth control literature unusable as a basis for scientific knowledge.[74] Therefore, in addition to women's life stories, birth control advocates took up the tools of social science to bolster their claims. In late 1929, Sanger authorized a systematic analysis of 10,000 cases drawn from her clinic, which was published in 1933.[75] In particular, that study sought to demonstrate to skeptics that contraceptive methods indeed prevented conception, and that different groups of women could use the methods effectively. Population scientists dismissed the personal narratives as too emotional; they dismissed these quantitative clinic data as methodologically flawed because of their connection to advocacy. As Raymond Pearl noted, there was a "need for more critical objective evidence, and less chatter about birth control" because "there exist[ed] almost nothing in the way of critical, objective, unbiased evaluation of the effectiveness of any or all contraceptive techniques, as *actually practiced* in the population." Instead, "exaggerated inferences" were "drawn from meager experience, statistically considered." Moreover, "nearly all the so-called evidence" came from those "interested in birth control propaganda," and "it is sound human instinct to look somewhat askance" at such "testimony, and to attach but little weight to any protestation of honesty or nobility of purpose which may accompany it."[76] Pearl and Notestein set about producing that appropriately objective statistical knowledge.

Contraceptive-Use Studies

From the perspective of population scientists and eugenicists, clinical studies did not address the fundamental questions of concern to them: What effect did contraceptive practice have on "large population aggregates considered as wholes"?[77] Would it exacerbate differential fertility? These two questions were always intertwined. Anxiety about the potentially negative effect of contraception on the size and quality of the national population circulated widely. But little "reliable" data existed to answer these questions. Therefore, in 1931, with the influence of Frank Notestein and Frederick Osborn, the MMF sponsored studies by Raymond Pearl and Frank Notestein designed to provide independent statistical evidence about contraception's effects.[78] Together these studies produced an account of US contraceptive practice based on statistical measurements of pregnancy risk, natural fertility, and contraceptive competency. This naturecultural[79] discourse secured the epistemological approach of fertility studies.[80] Notestein's analysis also took up the additional task of evaluating the contributions of the birth control clinic to patients' competence and delimited the proper scope and authority of birth control clinics. The recuperative effects of these measurement practices become clear when one focuses on the gendered assumptions underlying them against the backdrop of women's experiential knowledge as archived in birth control literature.

The Risk of Pregnancy

For Raymond Pearl, the leading biometric statistician in the United States, the "first step necessary to any scientific understanding of the effect of contraception" was to obtain "clear-cut" answers to three questions: To what extent "statistically" is contraception "actually practiced?" What was "the quantitative effectiveness" of the various contraceptives "both separately and all together in reducing the relative frequency of pregnancy, as they are actually used?" And what impact did contraception *"as actually practiced in the population"* have on aggregate pregnancy rates?[81] From his point of view, studies based on birth control clinic records were methodologically flawed, in part because, by looking only at contraceptive users who visited their clinics, they confused the "potentiality and the actuality of effectiveness." As he argued: "It is dubious logic to reason from the fact that a highly intelligent woman, thoroughly trained in biology in the university, and obsessed with an overwhelming fear of unwanted pregnancy is able to use a particular contraceptive device with unfailing success, to the conclusion that this contraceptive device is, or will be, equally effective as actually used by all women who resort to it in the general population. Nor can it be safely inferred from the same premise that birth control is a major factor in causing the decline in the birth rate."[82]

The purpose of his study was "to get direct, observed evidence of . . . the actual as distinguished from either the theoretical or potential effect of contraceptive practices upon the . . . overt expression of natural fertility." Pearl's formulation of the distinction of potential and actual effectiveness, as a distinction between the practices of "intelligent and careful persons" and "the general population," announces the elitist assumptions of eugenic fertility discourse that drove his research interests.[83] While highly intelligent and vigilant (i.e., white, middle-class) women might use contraception very effectively, to generalize from their experience would, he concluded, overstate the actual level of competent contraceptive practice in a "general population" of diverse character. The matter of contraceptive competence will be discussed in detail below. First, I look more closely at the statistical innovations he developed to distinguish the potential effects from the actual, that is, "the real," effects of contraception.

To answer his three research questions, Pearl supervised the collection of 5,000 case histories through a national network of physicians in large urban public hospitals who interviewed all the women admitted to the maternity ward for childbirth during 1931 and 1932. The sample included both contraceptive users and non-users. For each case, information was recorded on all pregnancy events and outcomes, contraceptive methods, and the patterns of their use. The data also recorded information about each woman's social identities: age, race, education, economic standing, and husband's occupation (figure 4.1). Pearl's 1932 preliminary analysis of these data yielded a "rough crude answer" to the first research question. With a straightforward calculation of simple percentages, he found that 29.55% of the women in the sample had actually practiced contraception.[84]

Answering the second and third questions was a more complicated matter, however. It required innovative means for calibrating the various observed pregnancy rates of groups of non-users and users of different contraceptives for different lengths of time. What was needed, he argued, was a standard metric of the probability of pregnancy, which would smooth out these differences. He constructed such a metric with the concept of the *duration of exposure to the risk of pregnancy.* Pearl started with the basic assumption that "a woman past puberty but not past the menopause is assumed to be exposed to the risk of becoming pregnant when she is more or less regularly indulging in sexual intercourse, as in the married state." From this, he concluded that for "practical purposes"

HOSPITAL

OBST. NO.

DATE OF DELIVERY

LEGIT. OR ILLEGIT.

DO NOT WRITE IN THIS SPACE

COLOR	RACE STOCK	RELIGION	HAS PATIENT ANY GYNECOLOGICAL DISEASE? IF SO SPECIFY.
W. C.			

REPRODUCTIVE HISTORY INCLUDING PRESENT ADMISSION

PREGNANCY	YEAR	RESULT
1		L. S. M. T. O.
2		L. S. M. T. O.
3		L. S. M. T. O.
4		L. S. M. T. O.
5		L. S. M. T. O.
6		L. S. M. T. O.
7		L. S. M. T. O.
8		L. S. M. T. O.
9		L. S. M. T. O.
10		L. S. M. T. O.
11		L. S. M. T. O.
12		L. S. M. T. O.
13		L. S. M. T. O.
14		L. S. M. T. O.
15		L. S. M. T. O.

L=LIVE BABY. S=STILL BORN.
M=SPONTANEOUS MISCARRIAGE
T=THERAPEUTIC ABORTION.
O=OTHER ABORTION

HAS PATIENT EVER USED ANY METHOD FOR PREVENTION OF CONCEPTION? YES. NO.
(FILL IN DETAILS ON OTHER SIDE OF CARD)

DATE OF BIRTH OF PATIENT?

DATE OF BIRTH OF HUSBAND?

DATE OF MARRIAGE?

WARD, PAY, OR PART PAY PATIENT

OCCUPATION OF HUSBAND?

EDUCATION OF PATIENT: ILLITERATE / ELEMENTARY SCHOOL / HIGH SCHOOL / COLLEGE

ECONOMIC POSITION: VERY POOR / POOR / MODERATE CIRCUMSTANCES / WELL-TO-DO / RICH

HAS PATIENT EVER HAD SELF-INDUCED ABORTION? YES. NO.
HAS PATIENT EVER HAD ABORTION INDUCED BY SOMEONE ELSE? YES. NO.
(IF ANSWER IS YES IN EITHER CASE, DESCRIBE METHOD USED)

NOTES:

THIS CARD WAS FILLED OUT BY:

OVER

History Card Used in the Investigation. Obverse

METHODS OF CONTRACEPTION USED
(TO BE FILLED IN WITH AS MUCH DETAIL AS POSSIBLE)

METHOD	CHECK USE	HOW LONG WAS EACH OF SPECIFIED METHODS PRACTISED?	WHAT IS PATIENT'S OPINION AS TO EFFECTIVENESS OF METHODS SHE HAS USED?
Coitus Interruptus (withdrawal)			
Condom { rubber / skin			
Pessary alone			
Pessary with medicated jelly			
Pessary with douche			
Medicated Vaginal Suppositories or Jellies*			
Douche alone—water			
Douche alone—medicated*			
Intra-Uterine Mechanical Device*			
"Safe Period" (abstinence during part of month)			
Any Other Method*			

* Specify Kind Here:

DO NOT WRITE IN THIS SPACE

OVER

History Card Used in the Investigation. Reverse

FIGURE 4.1. Raymond Pearl: case history card (front and back). From Pearl 1932, 370–371. Courtesy of Wayne State University Press.

exposure to the risk of pregnancy is equivalent to the length of time a woman has been married.[85] Of course, he reasoned, one could not get pregnant while pregnant, so the *duration of exposure to risk* required that time spent in pregnancy be subtracted from the total years of marriage. The pregnancy rate thus consisted of the ratio of the aggregate number of pregnancies to the aggregate *duration of exposure to risk of pregnancy* for each social group. Pearl concluded that his formulation of pregnancy risk increased the accuracy of statistical analyses of contraceptive effectiveness because it calibrated women's entire reproductive histories to a single scale that controlled differences of age, duration of marriage, and contraceptive types and use patterns.

Initially Pearl used the total number of years married but not pregnant, aggregated by calendar months, to calibrate risk exposure periods. By 1933, he had refined the risk measurement to get nearer to "the real biological roots of the matter." Instead of units of marital life while not pregnant, he incorporated a new standard for measuring pregnancy risk: ovulation. In Pearl's view, knowledge of the physiology of reproduction made it possible to construct an approximate measure that was "sufficiently close to the actuality" to warrant its use. Thus, he argued, "what we really want as a measure of pregnancy-rate is the answer for each individual woman, and constructively for the group, to the following question: What proportion of all the ovulations experienced during the period of observation resulted in the fertilization of an ovum or ova and pregnancy?"[86] The refined pregnancy rate was calculated as follows: The numerator equals the total number of times pregnant during marriage, multiplied by 100. The denominator equals the number of ovulations per year (13) multiplied by the total duration of marriage between puberty and menopause minus time spent pregnant, and that figure is added to the number of pregnancies. The mathematical equation is shown in figure 4.2.

"The addition of a constant multiplying factor for ovulation" had "the effect" of simplifying the math and providing a regularized unit of duration of risk. It also amplified the data points obtained from the sample because each ovulation became a discrete time unit of risk. From the 2,000 case histories, this method yielded 6,869 person years of exposure to risk. The 5,000 cases analyzed in Pearl's 1934 articles yielded 17,505 person years of exposure. The greater the person years of risk included in the statistical analysis, Pearl argued, the greater the reliability of the

We may then set up the pregnancy-rate as follows:

Let M = the total period (in years) during which a woman engages in copulation, between puberty and the menopause (for practical purposes the duration of marriage within the same limits—that is, between puberty and the menopause); and

P_1 = duration of time (in years) she spends in the pregnant state, and regardless of the manner of its termination (by term birth, or abortion, etc.); and

T = number of times she becomes pregnant during the time-period M; and

P_2 = duration of time (in years) she spends in the puerperal state (taken to a rough approximation as $.04T$ or $.04(T-1)$ when, as is the case in our records, the period of observation and record ends with the date of the termination of a pregnancy).

Then $M - P_1 - P_2$ = duration (in years) of exposure to risk of pregnancy, and

$$\frac{100\ T}{13\ (M - P_1 - P_2) + T}$$

$= R_p =$ pregnancy-rate per 100 ovulations.

FIGURE 4.2. Raymond Pearl: calculation of the pregnancy rate. From Pearl 1934a, 357. Courtesy of Wayne State University Press.

impact of contraceptive use on pregnancy rates and, by extension, population trends. With this method, Pearl was able to demonstrate statistically that contraception was effective. "Birth control as actually used . . . was associated with a really lower pregnancy rate" for individual women and in the aggregate.[87]

The addition of ovulation to the equation may have made the mathematics more manageable. However, I contend, it also had important conceptual effects that made gender politics more manageable. By making ovulation the variable of "fundamental importance" to human fertility, Pearl's model rendered heterosexual sex statistically irrelevant to calculation of the risk of pregnancy.[88] Pearl's risk metric abstracted the risk of pregnancy from the social processes of marital sexuality and the historical circumstances of life as it was lived and converted it into a naturalized event in abstract units of biological time.[89] Pearl's formula thus disconnects pregnancy from the social processes in which it occurs and recasts it as a universal, natural process, "the fertilization of an ovum." There are no cold distances, no rebuffed advances, no slips of restraint, no hurried copulations, or any tender feelings. There are no bodies, no fluids, no heat, and no emotion. In fact, there are no sperm—only ova, whose presence reflects the natural biological rhythms of women's bodies, independent of social practices. Ovulation events were not observed, however. They were conjectural events based on the model of normative menstruation charted in biological discourses in the 1920s and 1930s.[90]

Pearl's conceptualization of pregnancy risk inscribed the ongoing cultural shift in the burden of reproductive management. Prudential restraint had placed the burden of risk and its management on men. With the renegotiation of the normative terms of companionate marriage, this burden gradually shifted to women. This change is reflected in Pearl's analysis of contraceptive effectiveness. The 1932 preliminary report included a statistical analysis of which partner "take[s] the trouble" of contraceptive practice, indicating that in about half the sample men bore the burden and in the other half women "take all the trouble and do all the work of preventing conception." This conclusion was based on the rate of male-controlled methods (condoms and withdrawal) to the female-controlled methods (douches, foams, suppositories, and related vaginal methods). In the same 1934 articles in which he espoused the ovulation constant, Pearl collapsed all contraceptives into a single category, ignoring the gender specificity of methods and users. Instead, the sample situated women as fully responsible for contraceptive practice and arrayed them on a scale from "stupidity about contraception" to "intellectual enlightenment."[91] Within this analysis, pregnancy risk was configured as if women faced it alone, because of their nature, and thus women alone were accountable for its management.

Secured to ovulation, his representation of pregnancy risk closed down discursive space in which the gendered tensions in heterosexual relations might have been observed or theorized.[92] Women's experiential knowledge of the embodied social processes that defined pregnancy risk within the constraints of conventional femininity became illegible in Pearl's figuration of risk. Risk resided in women's bodies alone; it was no longer a product of marital sex. The complex negotiations that characterized women's descriptions of marital intimacy were thus also irrelevant to the evaluations of "*actual*" contraceptive practice. In this way, Pearl's measurement practices instantiated the emerging ideology of male-dominant companionate marriage—more sex, fewer kids—and suppressed the gender-based disjunctures between the normative narrative of heterosexual marriage and reproductive life as it was "lived in the individual" for modern women.[93]

Natural Fertility

Frank Notestein, who received his PhD in social statistics from Cornell University in 1927, joined the MMF as the director of its population division in 1928. There, together with research associate Regine Stix, he conducted a follow-up study of patients from Sanger's New York City clinic, published as *Controlled Fertility* in 1940. The goal of the study was "to exploit" Pearl's "superior methodology" for assessing contraceptive effectiveness using the clinic's comprehensive database.[94] It drew a sample from the patients who resided in the Bronx and attended the clinic between 1931 and 1932. The women were recruited to participate through a letter signed by Sanger. Regine Stix, who worked under Notestein's supervision, conducted one-hour interviews in the women's homes, during which she collected an exhaustive reproductive history of all conceptions, pregnancies, pregnancy outcomes, and contraceptive practices, as well as extensive data about their social position. Notestein then compiled a "careful evaluation" of the patients' reproductive history both before and after "clinic instruction." With this information, Notestein attempted "to evaluate the importance of factors" that contributed to successful contraceptive practice.[95]

The bulk of the analysis consisted of a reappraisal of the effectiveness of various methods using Pearl's pregnancy rate to correct the principal "methodological error" of prior studies produced by birth control clinics. Those studies relied on "case-failure ratios," a simple comparison of the percentage of pregnancies that occurred while using various contraceptives. The straight comparison of success/failure percentages was statistically inadequate, Notestein argued, because it did not account for different durations of exposure to the risk of pregnancy. That is, he argued, women likely used the prior methods for much longer periods than they used the clinic methods. Thus, the clinic method had an unfair advantage in case-failure comparisons because its effectiveness was measured for a shorter period of risk, a period in which only one pregnancy would have been expected in any case. Other methods were likely judged against time intervals in which multiple pregnancies would be expected. In case-failure calculations, the success rates of other methods would thus be low, not because they were "ineffective" but because the "trial period was long."[96]

Using Pearl's pregnancy rate formula, Notestein controlled for duration of exposure to risk when calculating relative contraceptive effectiveness (figure 4.3). The calculation relied on two data points: "(1) the number of pregnancies experienced" in the sample while using contraceptives; and "(2) the number of pregnancies that would have been experienced during the exposure had no contraception been practiced." The number of pregnancies that occurred was a straightforward calculation. It was then sorted by contraceptive type and aggregated into person months of marriage. The tricky part was the estimation of the number of pregnancies that would otherwise have occurred. Measuring these non-occurrences required ingenuity. No standard estimate existed of the "reproductive potentialities" of "uncontrolled fertility" among humans.[97]

Notestein produced such an estimate by building on Pearl's concept of pregnancy risk. Thus, ovulation was again conceptualized as the risk event. Notestein began by figuring out the sample's actual pregnancy rate during months in which contraception was not used. From that observed rate, he extrapolated an aggregate expected lifetime pregnancy rate for the sample and represented it "dramatically . . . as if it were that of an 'average woman' who made no attempt at contraception throughout her married life." That figure: 14 pregnancies between 20 and 45 years of age. This estimate of uncontrolled fertility "set up a control by which the effectiveness of contraceptive practice [in a population] could be measured" in terms of the reduction in "the risk of pregnancy in this group." That is, the difference between the actual number of pregnancies and the expected number was equivalent to the amount of aggregate risk reduction. Specifically, subtracting the actual number of preganancies with contraception from the number expected without it, Notestein obtained the number of prevented pregnancies. In turn, that number divided by the expected number represented the proportion of pregnancies prevented by a contraceptive, which he took as a measure of effective "contraceptive practice."[98]

The analysis presented in *Controlled Fertility,* thus depends on the co-configuration of measurements of uncontrolled (i.e., natural) and controlled fertility. The numeric assessment of controlled fertility was calibrated across time measured in ovulation months. It is a measure of deviation from a statistically configured aggregate measure of uncontrolled fertility. Uncontrolled fertility represents women's unrestrained

$$\text{Rate (Pregnancies per 100 years of exposure)}^{a} = \frac{\text{No. of pregnancies}}{\text{No. of years of exposure to risk}} \times 100$$

FIGURE 4.3. *Controlled Fertility:* the formula for pregnancy rate calibrated to risk. From Stix and Notestein 1940, 169. Courtesy of Wolters Kluwer | Lippincott, Williams and Wilkins.

biological (i.e., natural) reproductive capacity. The "as-if" average woman who never practices contraception and has 14 pregnancies in 25 years is imaginary, however. She is a manifestation of the measurement methodology that treats reproductive potentiality as an artifact of ovulation in marital time.

Among clinic-goers in New York during the depths of the Great Depression in 1931 and 1932, a large proportion of the months during which they did not use contraception were clustered in the period between their marriage and a first pregnancy. Even so, 40% of the women reported contraceptive use immediately after marriage. The number who reported contraceptive use after their first pregnancy increased the proportion of contraceptive users to 80%. This proportion approached 100% as the duration of marriage increased. Notestein estimated that women in the sample practiced contraception between 90 and 99% of the period of exposure to the risk of pregnancy.[99] This figure is unsurprising when one considers that the sample period involved one of the worst economic crises in modern times. But as a result, the observed episodes of "uncontrolled fertility" in the sample were very limited and occurred most among newly married couples. Therefore, *Controlled Fertility*'s estimate of the expected lifetime natural pregnancy rate, the "as-if" woman, effectively aggregated the contraceptive and sexual practice characteristic of newly married couples into a general pattern and extrapolated that across the reproductive life span. Early marriage, however, was a period in which women's narratives, Pearl's research and Notestein's own data suggest, saw higher levels of sexual activity, lower levels of contraceptive use, and lower levels of marital tension over fertility control.

Calculating the expected pregnancy rate based on the pattern of early marriage as an algebraic constant across the "as-if" woman's life span, Notestein produced a figure of 14 pregnancies for a lifetime of uncontrolled fertility. This figure was nearly double the rate of 7.8 births population scientists estimated for US women in 1790, "one of the most fertile [populations] ever observed."[100] This did not dissuade him from using it as the metric by which to interpret the meaning of the rates his analysis produced. In fact, Notestein concluded that his analysis might have underestimated the potential reproductivity of the group.[101] A high rate of uncontrolled fertility followed from the underlying Malthusian logic, in which only prudence and forethought forestalled nature's "superabundant effects."[102]

Notestein argued that calibrating contraceptive effectiveness against the constant high risk of uncontrolled, natural fertility provided better estimates than the simple case-failure method because it controlled for durational differences. With this measurement tool, he asserted, it became effectively possible to isolate the degree to which clinic attendance influenced contraceptive practice. The inferences he drew about the comparative effectiveness of different contraceptive methods, and for that matter the overall level of control, hinged on the accuracy and reasonableness of the calculation of expected pregnancies. If the approximation of the expected number of pregnancies in the absence of contraceptives overestimated the actual number of pregnancies that would otherwise occur, it would overestimate the number of prevented pregnancies and amplify the apparent effectiveness of a method. For instance, if the expected number of pregnancies was 10 and the observed actual number was 8, the formula would be $10 - 8 = {}^{2}/_{10} = .2$, or a 20% effectiveness rate. If on the other hand the expected number of pregnancies was 14 and the observed actual was 8, the formula would be $14 - 8 = {}^{6}/_{14} = .428$, or a 43% effectiveness rate. Thus, overestimation of uncontrolled fertility and prevented pregnancies could inflate the effectiveness rate of all methods while obscuring the differential effectiveness between them.

Judging from the far higher effectiveness rates *Controlled Fertility* reports compared to other studies, this appears to be what happened.[103] That is, Notestein found that methods women used before coming to the clinic, which he termed "untutored efforts at contraception," demonstrated a "surprisingly" "high degree of effectiveness," at 79% overall. (Condoms were calculated to have prevented 86% of expected pregnancies, coitus interruptus was 78%, and even the poorest method, douches, had a 61% effectiveness rate.)[104] The effectiveness rate for the clinic-prescribed method (the diaphragm) was highest, at 94%. But because the effectiveness of all methods improved post-clinic attendance, he concluded that the diaphragm, the clinic method, was not a significantly better method. The higher rate for the clinic method, he concluded, likely combined the effects of both method quality and tutored use. Moreover, he discounted the advantage gained by using the clinic's method, calculating that only 40% of the sample continued to use it exclusively.[105] He concluded that all methods except douches were reasonably effective, and the differences between them were minimal. "Untutored practices," "which have come down generation to generation time out of mind," were it seemed "highly effective" in the aggregate. In fact, the results showed that "even the use of relatively ineffective methods brings substantial reduction in fertility [rates]."[106]

Controlled Fertility's analysis of contraceptive effects involves an important shift of perspective from the individual to the aggregate. The key to contraceptive effectiveness from the aggregate perspective is the ability of contraceptive practice to alter group fertility trends, not the ability to prevent *any* pregnancy in an individual case. When one considers that the high effectiveness of pre-clinic methods that Notestein praised were the same methods that women sought to replace, it is clear that what counted as highly effective meant different things to them. The increase in post-clinic effectiveness of all methods—about 90%—surprised Notestein, but at least he noted that women who went to the clinic "in search of effective contraception found it."[107] Still, the shift of perspective in the statistical measurement of effectiveness removed the evaluation of contraceptive quality from life-as-it-is-lived, from the specific social historical situation of its practice, both the context of marital sexual relations and the specter of the loss of a baby's or one's own life. In its place, the women who populate the analysis are "as-if" women, averaged figures, mapped onto biological time (100 person years of ovulation) for whom all methods are adequate in the aggregate. This "as-if" figure displaces the fleshy world of inadequate access and uncertain effect that women narrated in birth control literature.

Also, as in Pearl's analysis, the use of ovulation as the measure of risk invested reproductivity in the natural regularity of female biology. Although Notestein observed that "obviously one of the variables to be considered in comparing risks is frequency of coitus," he concluded that it did not have much impact on the average number of months before a pregnancy occurs.[108] In addition, although he noted that "marital adjustment" and "satisfactory coitus" are important linked factors leading women to the clinic, beyond a cursory investigation of coital frequency, specific sexual practices did not matter in *Controlled Fertility.*[109] In fact, marital abstinence was not assessed as a contraceptive method.[110]

Although largely ignored, glimpses of women's individual perspectives on marital sexual bargaining appear in *Controlled Fertility.* For instance, one chart quantified the "reactions of couples" to the three major methods—the condom, the diaphragm, and withdrawal.[111] The greatest degree of consensus was with the diaphragm; 49.6% of reports indicated that "both liked." The next strongest point of consensus was that condoms and withdrawal were "disliked" by both (39.6 and 39.8, respectively). The chart (figure 4.4) also included suggestive information about marital tensions in the percentages of the reports indicating conflicting reactions, where one liked and one disliked the method. The gendering of this conflict pattern is suggestive. Nineteen percent of wives who liked condoms and withdrawal, respectively, reported that their husbands were indifferent or disliked them, while 27.1% of women disliked the diaphragm and reported that their husbands were indifferent or liked it. This gender-specific source of marital tension was ignored. Gender-based tensions were also elided in the observation that "75 percent of the husbands preferred to have their wives assume this responsibility."[112]

Thus, like Pearl, by conceptualizing pregnancy risk as a consequence of ovulation, *Controlled Fertility* closed down discursive space within which gender-based tensions over marital sex and contraceptive practice might have been articulated, even as it affirmed the displacement of the Malthusian ideal of masculine prudential restraint with feminine contraceptive competency. The categories of Notestein's analysis left little room in which to examine how access, use, and satisfaction might be specified differently from the perspectives of women and men. Instead, by defining all existing contraception as adequately available and effective based on reduced pregnancy rates in the aggregate, *Controlled Fertility* could not register the marital tensions, inadequate access, and uncertain effects that women narrated in the birth control literature. Yet, as women's narratives suggested, gender differences mattered here. In the gap between companionate marriage ideals and lived intimacy, questions of access,

Reactions of Couples to Three Types of Contraception

Couples Ever Using	Total Number for Whom Reactions of Both Husband and Wife Were Known	Per Cent of Couples Reacting in Each Way						
		Total	Both Liked	Both Indifferent	Both Disliked	Wife Liked, Husband Indifferent or Disliked	Husband Liked, Wife Indifferent or Disliked	One Indifferent, Other Disliked
Clinic Prescription..........	676	100.1	49.6	3.0	9.8	6.5	27.1	4.1
Condom....................	502	100.0	11.6	3.6	39.6	19.7	7.4	18.1
Coitus Interruptus...........	387	100.1	7.8	4.7	39.8	19.1	9.6	19.1

FIGURE 4.4. *Controlled Fertility:* reactions of couples to three types of contraceptive. From Stix and Notestein 1940, 101. Courtesy of Wolters Kluwer | Lippincott, Williams and Wilkins.

agentic use, and satisfaction were very different for diaphragm, condom, and withdrawal. Once they were obtained, diaphragms did not have to be persuaded to cooperate. The technique of condoms and withdrawal might have been readily available to married women in theory, but their practice was far more complicated, as the Sanger and Florence texts testify. The diaphragm may have been a difficult technology to manipulate for a generation of women without a lot of hands-on experience with vaginal barrier methods, but withdrawal and condoms often required their own complex negotiations and uncertainties. Instead, for Notestein, since "most methods were sufficiently effective," the important factors shaping contraceptive practice were "the attributes" of the users.[113] Disconnected from the fleshy world and structure of feeling in which contraception is acquired and used, the demographic measurements of motivation would simply distribute contraceptive practice across a raced and classed hierarchy of culturally (in)competent femininity that renders racial privilege and economic advantage invisible.

(In)Competent Contraceptors

The universalized female figure of uncontrolled fertility is only one of the figures that populates the landscape of Pearl's studies and *Controlled*

Fertility. She is joined by the figures of controlled fertility: the competent contraceptive user and her more "careless" sisters, who represent class and race difference. Recall that Pearl's innovative risk measurement was developed in the service of his investigation of differential fertility. Likewise, as Notestein argued, "it is not enough to know that contraception can be, and in certain instances is, effectively employed to prevent conception. It is also necessary to know the extent to which various groups use it, and the effectiveness with which they do so."[114]

To get at the question of the underlying causes of differential fertility, Pearl's studies distributed cases on the basis of two intertwined schemes: social characteristics of users (class, race, ethnicity, and religion) and personal characteristics of use (consistent users, careless users, and non-users). In the 1932 article, Pearl reported data only for race differences, and did so in terms of very disparaging stereotypes. In particular, he relied on the canard that "the American negro . . . exercises less prudence and foresight than white people do in all sexual matters."[115] Yet, overall, the data also surprised him. The differential pregnancy rates were quite small. Furthermore, the data indicated that birth control did not exacerbate differential fertility. His 1934 analyses demonstrated that the pattern of pregnancy differentials among users and non-users was essentially identical in shape and trend, which, he noted, was "somewhat surprising." Those who used contraceptives with consistency and planning in every class and race experienced the same relative decrease in pregnancy rates. Even those who used contraceptives intermittently had lower pregnancy rates than those who did not use contraception. Thus, biological differences could not account for differential fertility rates.[116]

Still, where biology alone failed to sustain difference claims, Pearl's typology of contraceptive practice provided a substitute variable. His 1932 preliminary analysis simply categorized women into users and non-users, but he found that this produced too diverse a group to allow for effective statistical analysis of difference. His 1934 analysis divided the user category in two: those who were steady users and those who were intermittent users. He then analyzed the patterns of contraceptive practice by two factors, *genus* (the degree to which methods were practiced with consistency and intelligence) and *species* (the method or methods utilized). The goal was to produce categories that could measure "the effect of intelligent and precisely performed contraception upon pregnancy and birth rates."[117] What is important about these figures is the way in which Pearl's assumptions about differences between classes and races are configured within them.

He began with the assumption that "the frequency of the practice of contraception is presumably in some degree definitely correlated with the degree of general enlightenment in any population." Consistent users were then defined as those who "did their contraception intelligently, precisely, and 'effectively,'" where effectiveness was defined as 100% success in preventing unwanted pregnancies.[118] The category of intermittent use, on the other hand, was differentiated by motive. Those who reported intermittent use because they planned to become pregnant were grouped with the steady users. Otherwise, the intermittent user group represented "carelessness."[119] In fact, non-use for a planned pregnancy was the only kind of intelligent non-use that was legible in Pearl's categorization. Thus, instances of intermittent use attributed to dissatisfaction with or difficulty in using the method were simply classified as instances of carelessness. In this way, women's dissatisfaction with contraceptive methods was converted into a personal failing. Intermittent users and those who used contraception regularly but experienced a pregnancy anyway were subject to pejorative judgments by Pearl, who described them as "feckless." He characterized them as suffering from "a combination of bad luck and bad management," but concluded that "actually most of the 'bad luck' was probably really bad management, due to the general ignorance and lack of specific understanding of the biological principles involved in human reproduction."[120]

On the surface, these differences are not race and class specific; "under the thralldom of the sexual urge," even "the economically better off folk contain among their number a residue of about the same proportion of people who, though well intentioned about contraception, are reckless upon occasion." However, he noted with high praise that "well-to-do" white women were more likely to be consistent, "sharp-witted and intelligent in their contraceptive practice."[121] His most pejorative characterizations refer to African Americans, whose "appallingly high" proportion of contraceptive failure he attributes to "lack of intelligence."[122] Such evaluations illustrate the underlying class and race assumptions about competence that Pearl's analysis relied on and reproduced. His evaluation of the relative level of enlightenment among different social classes and racial groups discounted questions of differential cost and access to methods. For instance, he recognized that "middle-class" women relied most on condoms, while women of the lower economic classes and African Americans were more likely to rely on douches and withdrawal. He noted, in fact, that "economic position altered contraceptive practice . . . the percentage using condoms [rose] steadily with advancing economic status." He also noted that condoms were the best method available to most women. However, he attributed the higher proportion of white well-to-do condom users to their greater understanding and appreciation of its reliability, discounting cost as an element of method selection. Instead, these differential patterns of method and consistency of use were configured as evidence of differential competence. Such qualities were, he concluded, "matters of character, bred in the bone and unlikely to be rapidly altered by experience."[123]

In a similar line of argument, the competent user of *Controlled Fertility* was the averaged white-collar worker's wife who sought contraceptive advice on her "own initiative" and used it continuously and exclusively. Like Pearl's potential user, she was smart, highly motivated, and careful in her habits. The analysis showed that middle-class professionals had the lowest fertility rate, the highest usage rate, and the most effective practice, both in terms of the number of accidental pregnancies and the selection of the most effective methods. When pregnancies occurred in this group, they were categorized as planned, based on women's reports that they had ceased using contraception because they "wished" to become pregnant. Thus, *Controlled Fertility* concluded, "the fertility of the white-collar workers was lower than that of the manual workers, largely because the white-collar workers used contraception more effectively." In Notestein's as in Pearl's analysis, failure resulted from human, not technical, inadequacies or the lack of access to "something certain." While Notestein acknowledged that effectiveness was always a combination of means and motivation, technology and techniques, tools and practice, the aggregate effectiveness of the methods suggested that individual failures likely resulted from characterological deficiencies. Individual failures signified carelessness, imprecision, and ignorance. The patterns of individual failures, in turn, added up to characterological deficiencies of class and education. Notestein concluded: "With discouraging uniformity, the manual workers and the uneducated made less effective use of contraception than the white-collar class and the educated. This relation appeared in the selection of methods and in the effectiveness of their use."[124]

As in Pearl's analysis, Notestein's discounted suggestive evidence about differential economic access to contraception. The conclusion was simply that class differences in contraceptive practice represented differences in character and competence. In fact, the analysis included an explicit measure of character and competence in a table that arrayed the frequency of accidental pregnancies by the metric of neatness and precision (figure 4.5). A footnote to the table explains that the interviewer used a three-point scale to rate the personal appearance of each interviewee and her home. The scale consists of the categories careless and slovenly, average, and neat and precise appearance. Not surprisingly, the frequency distribution of accidental pregnancy by this assessment of appearance tracked that of reported class and education. Together these data configured competent contraceptors as women who were clean and careful as well as better educated members of the middle class.[125] Such women were represented as more accustomed to the pace, values, and promise of urban industrial life. Fewer such women were said to be found among the poor and uneducated.

However, again, these figures did not represent actual women. They were another version of "as-if" women, an aggregate figuration that tied color, class, and competence together. That is, Pearl's and Notestein's measurement practices disconnected information about pregnancies and contraceptive practice from the historical context and social relations of individual women's lives as they were lived, converting them into discrete events. The mean pregnancy rate and composite contraceptive-use patterns were then aggregated in an abstract typology of race and class categories that relocated difference in assumptions about morality and character. The resulting numeric value described the social character of each group on a scale of prudent reproductive performance calibrated to the careful and consistent contraceptive practices of women. Early demographic measurements thus recognized the similarities of fertility rates and the impact of contraception on those rates, and, at the same time, reaffirmed a cultural differentialist hierarchy of competent, prudent reproductivity. Pearl's typology of users, especially the logic that assembled the category of intermittent users, displaced eugenic biodeterminist

Pregnancy Rates with Clinically Prescribed Contraceptives, by Neatness Rating

Period of Married Life	Neatness Rating[1]			
	Total	Careless and Slovenly	Average	Neat and Precise
	Pregnancies per 100 Years of Exposure			
First Pregnancy	8	—[2]	11	6
All Later Pregnancies:				
Total	9	12	9	7
(Years since Marriage)				
0–4	13	34	12	7
5–9	12	13	13	8
10–14	5	13	4	3
15–29	6	3	6	7

Period of Married Life	Number of Years of Exposure and Number of Pregnancies							
	Total		Careless and Slovenly		Average		Neat and Precise	
	Exp. Years	No. of Preg.	Exp. Years	No. of Preg.	Exp. Years	No. of Preg.	Exp. Years	No. of Preg.
First Pregnancy	38.4	3	3.2	0	17.7	2	17.6	1
All Later Pregnancies:								
Total	701.9	65	122.1	15	413.4	39	166.4	11
(Years since Marriage)								
0–4	150.6	19	11.7	4	98.7	12	40.2	3
5–9	249.6	30	47.5	6	142.5	19	59.6	5
10–14	193.6	10	30.7	4	126.0	5	36.9	1
15–29	108.1	6	32.2	1	46.2	3	29.7	2

[1] The rating of the interviewer, at the time the record was made: 58 per cent of the women interviewed were rated as average, 22 per cent as unusually neat, and 20 per cent as slovenly and careless. The rating depended on the condition of the home and the personal appearance of the woman interviewed and of her children.

[2] Less than 10 years of exposure.

FIGURE 4.5. *Controlled Fertility:* pregnancy rates by neatness rating. From Stix and Notestein 1940, 116. Courtesy of Wolters Kluwer | Lippincott, Williams and Wilkins.

logic while extending Malthusian self-mastery to feminine contraceptive practice.[126] Notestein's configuration of education, class, and personal habits, likewise, gave numeric credibility to the bourgeois discourse of reproductive self-mastery that scaled differences to competent contraceptive performances.[127]

Pearl's and Notestein's figures of risk and control augured the (racialized and classed) gender figures of careless and careful contraceptors who populated the mid-twentieth-century population imaginary. These "as-if" women became the normative demographic figures through which fertility and its control became legible and through which race, class, and cultural differences were and are understood. Pearl's assessment of pregnancy risk and contraceptive effectiveness would come to animate global population control discourses, a future indicated in the final paragraphs of his 1934 article "Contraception and Fertility in 4945 Married Women," where he noted that the finite size of the "earthly globe" suggested a greater need for "man's control of nature and of himself."[128] *Controlled Fertility* also pointed to the future concern with global population growth in the observation that "more than half of the world's population is living in areas where the high ratios of people to developed resources perpetuate the cycle of high fertility, poverty, and human suffering."[129] There was some reason for optimism, Notestein concluded, because "although this group [manual laborers and the uneducated] used contraception less effectively than others, they nevertheless used it with a fair degree of absolute effectiveness." Differential fertility was likely to persist, however, which suggested an appropriate role for birth control advocacy.[130]

The Proper Field of Birth Control

Raymond Pearl and Frank Notestein both initiated their research at the same moment that two key analyses of clinic data were being carried out under birth control movement auspices.[131] As noted above, Pearl and Notestein specifically positioned their analyses in contrast to these movement-based studies. An explicit goal of *Controlled Fertility* was also to reappraise the value of birth control clinics themselves, to assess their "influence" and thus demarcate their "proper field" of action.[132] The opening chapter framed the matter as follows: "If a clinic is to justify its existence, the contraception practiced by its patients must be appreciably more satisfactory than that practiced before clinic instruction. Moreover, there must be reason to suppose that a substantial part of the improvement following attendance is due to the training received. It must not be wholly accounted for by such extraneous factors as the increased determination to control fertility that prompted the patient to attend the clinic and at the same time led to increased diligence in her contraceptive efforts."[133]

Based on these criteria, *Controlled Fertility* concluded that although "casual speculation might lead to the conclusion that people seek the advice of birth control clinics because they are ignorant of contraceptive methods," in fact, "the converse is true." The data showed that a high percentage of women practiced some form of contraception before attending the clinic (96%). Furthermore, although the post-clinic contraceptive practice was more effective than the pre-clinic practice, the margin of improvement was small.[134] And while the small margin of improvement might suggest the value of clinic instruction, in his view, it could just as easily have been the result of stronger motivation to control fertility. To Notestein, the willingness to seek clinic advice itself suggested that clinic attendees had greater determination than the "general population."[135] Thus, although both the clinic method and clinic-goers were more effective, *Controlled Fertility* concluded, clinics probably did not have a major influence on the patient's contraceptive practice. In addition, he argued, change in aggregate rates was what mattered most, and the birth control movement had not caused birth rates to fall. Rates "started to drop, through the instrumentality of contraceptives that required none or simple equipment, long before there were any clinics of consequence." Moreover, the "aggregate number of [clinic] patients, although absolutely large, [was] relatively too small to have an appreciable effect on population trends." Nor had clinics provided the means by which the "rapid growth in Western populations had been checked." Instead, he argued, "the increasing use of such folkway methods [withdrawal], extensively supplemented in recent decades by more effective commercial ones, could fully account for the downward course of fertility." Therefore, "clinics have played a minor part" in the "direct dissemination of contraceptive knowledge to the public."[136]

The conclusion that the birth control movement had only a negligible effect on the proliferation of contraceptive practice followed a logic that would become the settled truth of the midcentury population imaginary. It follows from Notestein's assertion that "birth control did not spring suddenly into existence to implement an age-old desire to limit procreation. A strong current of causal sequence ran in the opposite direction. New patterns of living and new values brought growing interest in family limitation that spread the use of known methods and stimulated the development of new ones. In a real sense, modern birth control is as much a result of new interest in family limitation as its cause."[137]

In support of this claim, Notestein referred to the authority of the 1936 study by Norman Himes, funded by the MMF and published as *The Medical History of Contraception,* which he noted "[had] shown contraception to be as old as civilization itself."[138] Himes's analysis, which itself drew on Pearl's and Notestein's research,[139] concluded that "contraception, as only one form of population control, is a social practice of much greater historical antiquity, greater cultural and geographic universality than commonly supposed by medical and social historians. Contraception has existed in some form throughout the entire range of social evolution, that is, for at least several thousand years. The *desire for,* as distinct from the *achievement of,* reliable contraception has been characteristic of many societies widely removed in time and place."[140] Himes thus turned a wide spectrum of control, ranging from unconscious desire to conscious practice, into a single universal scale on which motives mattered more than means in shaping outcomes.[141] Based on Himes, and his own assessment that withdrawal was reasonably effective in the aggregate, Notestein concluded that the high fertility levels of earlier times simply meant that what must have been widely known "folkway methods" were not widely used.[142]

If access to new knowledge and more certain techniques did not account for the spread of modern contraceptive practice, what is the explanation? The will to consciously control fertility was an effect of the forces of modern life. Increased interest and diligence were functions of general social evolution.[143] Pearl's, Himes's, and Notestein's logic drew on the relatively new anthropological concept of culture—the "entire stock of material objects, techniques, customs, attitudes, and institutions"—that organized individuals into social groups and accounted for differences in forms of social action between different societies. Within this culture-based theory of society, social evolution resulted from the process of diffusion. That is, new (and implicitly better) practices, beliefs, and ideas moved outward from sites of innovation, where they were initially concentrated. Changes in social practices occurred first because of changes in material culture, especially science and industry, which drove the development of new technologies and techniques. Cultural beliefs, practices, and institutions, particularly religion, family, and community mores, which were more rooted in the hearts and minds of men, were said to lag behind. According to this "cultural lag" theory, new motives and practices appeared first in the modern industrial and managerial classes, who were proximate to the engines of innovation and progress, and diffused outward and downward to the masses.[144] The more remote and distant from the material forces of progress, the more cultural heritage and traditional practices slowed the ability of classes, races, and nations to take up modern innovations. Thus, cultural heritage constituted a drag on progress.

Applied to population change, as Notestein described it, the modern era brought with it an "unparalleled period" of population change that began "when civil order replaced social and political chaos" and let loose the "undreamed-of productivity" of "the industrial revolution" in Europe. The resulting "rising levels of living" first lowered mortality. Thereafter, "the entire drift of our urban industrial civilization . . . created new demands for the means of controlling fertility."[145] Specific factors spurring the proliferation of contraceptive practice included the Malthusian classics—the rise of individualism, increased materialism, the scientific spirit, and the higher cost of raising children. Persistent differential fertility rates reflected the lag as modern motives diffused out and down from centers of progress and innovation.[146] Although Notestein would not name it until after World War II, his narrative of population change in *Controlled Fertility* prefigured the differentialist logic of demographic transition theory, which indexes social evolution by fertility rates.

Given this narrative of population dynamics, it was a "sociologically-naïve view" to consider "agitation" to be a sufficient "causative factor" in "the diffusion" of contraceptive practice. Rather, at best, movement agitation was a "catalytic agent."[147] For Notestein, the birth control movement was simply "an agency pushing us down the mainstream of our cultural development."[148] That is, the movement's promise of superior knowledge and methods might beckon to women, but only the abstract social factors that motivated their interest in and determination to develop contraceptive competency could bring them through the clinic door. Notestein maintained this view throughout his career, noting in his 1982 retrospective on the field, "One might as easily say that enhanced motivation to restrict fertility was the cause of modern methods as say that modern methods were the cause of reduced fertility." This was "evident to everyone as soon as the rubbish was swept away."[149]

Notwithstanding its minor causal role, Notestein concluded that the birth control movement clinics "ha[d] and will continue to have other important functions." Their "educational services [had] hastened the growing interest in family limitation," and their clinics had been important sites of research and education in new and more effective contraceptives. Nevertheless, Notestein advised, the movement should realign its focus and priorities. "As at present operated, birth control clinics are more important as agencies for the promotion of health than as agencies of population control." But if their programs were properly focused on the "lagging spread" of "new values and new interests," clinics might speed the "'top' down" diffusion of "the small family pattern," especially to those "needing it most, the poor and the uneducated."[150] Moreover, Notestein also cautioned, "the movement was too single-mindedly occupied with the limitation of fertility." The movement should give more attention to "the positive aspects of birth control" and "new and wise incentives" to promote parenthood, because "the freedom to be fertile is as important as the freedom to limit fertility in a democratic society that intends to maintain its culture and its stock through a voluntary acceptance of the obligations of parenthood." Such an approach might also help reduce the proportion of the next generation produced by the poor and uneducated classes. In this way, the birth control movement could "make important contributions."[151]

Although he used the language of freedom, none of Notestein's advice for the birth control movement touched on women's concern with managing femininity within the gendered conventions of marital intimacy that made them responsible for its maintenance and vulnerable to its loss.[152] Women, as individuals, were not his concern. He was only interested in the aggregate. His advice resonated quite strongly with eugenic concerns about the size and quality of the "American" stock, particularly his prescription for who needed what kind of advice. In this sense, women were merely the objective access point for acquiring the pertinent information about the fertility control practices of other people.[153]

The recuperative effect of *Controlled Fertility* is also remarkable when read with an eye toward the underlying contest with the birth control movement. This text and the web of texts Notestein drew from and contributed to established the scientific ground upon which demographers captured authority for fertility matters. Notestein's publications, in particular, shifted the question of fertility control from an issue of practical access, the birth control movement's chief concern, to an issue of motivation. If practical access to reliable contraception were the main barrier to effective fertility control, then the birth control movement should be the authoritative voice on matters of fertility control, and its clinics should be funded. *Controlled Fertility* demonstrated, however, that socially produced motivation was the key issue with regard to competent contraceptive practice. Given the "facts" of natural fertility, pregnancy risk, and differential contraceptive competence, the first priority had to be scientific investigation of the social and psychological factors motivating competent contraceptive practice. Mere provision of contraceptives was not enough. Thus, population science was properly the authoritative voice on matters of fertility control, and its research deserved funding.

This logic defined the stance of demographers and funders toward birth control for the next generation. The performative effect of the text installed motivation as the key element in fertility control, and that motivation was grounded in modernity. In their proper field, birth control clinics were important educational and product-testing sites, but at a distance from and subordinate to the authority of population science. This logic also inscribed the eugenic concern with differential fertility into the next generation of research and the emerging population control establishment. While the causes were now seen as social rather than biological, these studies showed that the quantitative statistics of biometrics, which measured differences between social, ethnic, and national groups, could be used to measure the rates of diffusion and lag of cultural competencies. Abstracted from the conditions of life-as-it-is-lived onto the universal Eurocentric scale of progress, the "as-if" woman of natural fertility and her contraceptively incompetent sisters became world travelers. They became objects of the mid-century "bio-politics scripts" written into the demographic transition.[154]

• • •

Popular discourse often characterizes contraception as allowing greater sexual freedom because it separates sex from reproduction. This chapter shows another sense in which sex and reproduction were separated in the contraceptive age. The network of texts within the discursive community that had coalesced into the discipline of demography by 1940 produced modern population knowledge and standard demographic measurement practices that separated knowledge about reproduction and its control from the social processes of male-dominant heterosexuality. Demographic figures decontextualized fertility control from the structure of feeling, the interpersonal relationships, and the mortality of bodies within which women articulated their perspectives on reproductive risk. The core of women's concerns conveyed in letters and movement narratives—tensions over marital sexual relations and well-being constrained by local sociohistorical circumstance—was occluded within statistical figurations of pregnancy risk. In the demographic naturecultural model of risk, there is no keeping away from husbands, no tensions, quarrels, or threats to go elsewhere. There are no relenting indulgences, lapses of self-control, or anxious waiting for proof from month to month. There is no descent into the valley of the shadow of death, nor anguish over women and infants who die too soon. There is only the regular rhythm of ovulation, a function of normalized biological time. Thus, even as midcentury demography focused on events of women's bodies, it abstracted the facts of contraceptive use and effect from the messiness of life-as-it-is-lived and lost. Demographic measurement practices thus closed down discursive space in which the gendered terms of marital heterosexuality might have been theorized and contested.

The sentimental framing of women's narratives of contraceptive need and failure made them easy to dismiss within an ideological moment that valued the calm distance of scientism. Likewise, Sanger's call to harness emotion to promote social change was also easily dismissed as propagandist rhetoric. With their dispassionate metrics, demographic studies of metropolitan women's contraceptive practice discounted the significance of the women-led birth control movement in terms of its practical impact, its narrative of contraceptive need, and its epistemic authority. They shifted the focus of inquiry from practical access to *something certain* to abstract social-psychological motivations and aggregate effects. Thereafter, foundation-funded research focused on the cultural habits and values that shaped individual motivation to control fertility and thereby shaped aggregate fertility patterns.

But sentiment is not absent from demographic figures. The familiar Malthusian affect—anxiety about the excess fertility of others—resides in the co-configuration of uncontrolled and controlled fertility. That fear is instantiated in demographic figures of natural fertility, the "as-if" woman against which to measure fertility control. The product of the aggregation methodology itself, she is an imaginary woman, a specter who haunts the modern population imaginary: the woman who makes no effort at contraception across her reproductive life span, regardless of the consequences for society. Her effects are not even modulated by mortality: the "as-if" woman does not die in childbirth and her children do not die in infancy.[155] Mortality, which was central to Malthusian judgments, disappears. In its place fertility became the index of differences between men and peoples. Although these foundational studies displaced biology as the cause of differential fertility, they confirmed the neo-Malthusian logic that naturally excessive reproduction was only contained by enlightened practices. Since all contraception was good enough to reduce aggregate rates, the "as-if" woman could be used to calibrate the scale of contraceptive competence of various others. Social groups and nations could thus be characterized by the care or carelessness of the contraceptive practices of "their" women.

Neo-Malthusianism, therefore, was not simply a matter of authorizing contraception. It re-secured hegemonic masculinity by closing down discursive space in which negotiations of heterosexual expression might take place and by reframing the problem of population as a consequence of the ill-considered actions of racial, economic, and colonial others. The standard estimate of "natural fertility,"[156] developed in these early contraceptive-use studies, defined the intellectual grounds of midcentury demographic knowledge of the population explosion and justified "birth control for the nation."[157] Demographic conceptualization and measurement of (un)controlled fertility grounded the simultaneously empowering and disempowering effects of the modern contraceptive regime that targeted women's reproductive bodies, not as "Sanger's . . . autonomous sexual beings," but as actionable objects, "people something should 'be done to.'"[158]

Finally, while most disciplinary histories located the origins of demographic theory in the postwar period, this chapter demonstrates that postwar theory is prefigured in the contraceptive-use studies of the 1930s. The binary figures of the competent, enlightened modern women and the rural, backward, and traditional women who would be the objects of population control measures were constructed in *Controlled Fertility* and the web of texts that anchored it.[159] The next chapter examines the travels of the "as-if" women of demographic figures through the midcentury expansion of US social science to encompass the globe. It traces the social and textual articulation of the demographic transition, illuminating demography's significant contributions to the population control establishment and the mid-twentieth-century formation of hegemonic masculinity.

5

Demographic Transitions and Modern Masculinities

> Numbers . . . were part of a language of policy debate, in which their referential status quickly becomes far less important than their discursive importance in supporting or subverting various classificatory moves and the policy arguments based on them.
>
> —ARJUN APPADURAI

THE CONTRACEPTIVE EFFECTIVENESS STUDIES DISCUSSED IN THE previous chapter circulated within both the national and the international gentlemen's club of demography. Their interest in contraception sprang from a perennial concern about the threat posed by the differential growth rate of peoples and nations. In modern international affairs, the size of a nation's population and of its territorial reach signified its political weight in the world, while mortality and fertility rates distinguished the vitality of "civilized" from the lassitude of "backward" nations. In the waning days of colonialism, focus shifted from the absolute size of national populations to population density, the ratio of bodies to territory. While the question of population quality was never far removed from questions about quantity, in the 1930s, high population density raised the specter of expansionism, which could potentially produce "explosive" political situations.[1] This raised the specter of another world war and political instability in colonial regions. Demographic research sought to discover the causes of differential population growth, identify the consequences of such growth, and provide prescriptions for defusing that potential. New statistical tools promised to make more mathematically precise estimations of current and future population patterns possible. However, "scanty" data remained a significant problem, especially in regions of the Global South.[2] But by the close of the Second World War, innovations in demographic theory, data gathering, and analysis that compensated for incomplete data lent greater scientific credibility to demographic numbers. Demography had "become science rather than literature."[3]

In many ways, the postwar discourse of population problems rearticulated longstanding Malthusian gendered imperial logic. Yet, mid-twentieth-century demographic theory and resulting measurement practices put them together in new ways. Of greatest importance was the introduction of demographic transition theory, which conceptualized population change as an automatic feature of human social evolution. This theory explained the "unprecedented" "growth" of Europe and its settler nations, the apparent cessation of that growth in the early twentieth century, and the potential for "explosive" growth in the rest of the world. Moreover, it shifted population discourse from Malthus's preoccupation with checks on growth that operated through mortality—misery, war, and famine—to checks on growth through control of aggregate fertility rates. With this shift in focus, the question of how best to manage the natural excess of female bodies became central to population policy. At the same time, by elaborating "the interrelated nature of social, economic, . . . political change" from "the demographic point of view," demographers resituated population as a metric of colonial difference and themselves as vital participants in mid-twentieth-century population and development programs.[4]

Historians have identified the influence that prominent individuals such as John D. Rockefeller III, General William Draper, and Hugh Moore had on US population policy and international family planning programs.[5] However, the focus on these famous men obscures the extensive transnational network, sustained by American demographers, that created modern population figures and population control policy prescriptions. A small group of social scientists led by Frank Notestein and Frederick Osborn secured that network. Although publicly they "eschew[ed] the role of advocate,"[6] they were both gifted entrepreneurs who skillfully drew financial support and scientific credibility to their enterprises.[7] Together these two individuals and the organizations they led, the Office of Population Research (OPR) at Princeton University and the Population Council, had an enormous influence over postwar demographic knowledge building.[8] In these sites, the gentlemen's club of demography collaborated in creating a naturecultural world in which "the numbers worked" to sustain the demographic transition as truth discourse and as a social fact.[9] Many of the demographers who participated in the contraceptive revolution and population control establishment started their careers as fellows supported by the OPR and the Population Council. Disciplinary narratives recognize the importance of the network of demographers and funding configured by Notestein and Osborn and centered in the OPR and Population Council, but they discount the efforts of these men in building the population control establishment, arguing instead that demography was captured by the family planning establishment.[10] But, as this chapter demonstrates, demographers were not outsiders caught up in the family planning craze. They were successful competitors in the contest for dominance in meaning-making about population, fertility, and the future, and thus they were instrumental in building the international population control establishment.

The chapter juxtaposes a close reading of demographic transition theory in the published works of its champions, Frank Notestein and Kingsley Davis, with an analysis of the multidirectional transnational flows of that theory through demographic, policy, and popular sites. It thereby elucidates the interconnected epistemic lenses and social worlds through which transnational audiences of politicians, policymakers, and the "educated" and "informed public" were instructed to view the world as already overpopulated and to locate themselves within the matrix of population control solutions.[11] The analysis stresses the affiliations of the demographic texts and personnel with the gendered coloniality of modernization theory.[12] Reading demographic texts through the lens of hegemonic masculinity brings into particularly sharp relief the intertwined logics of gender difference and hierarchy that aligned the population crisis with Cold War social science and international politics. In this case, the fate of men and nations hinged on aggregate national fertility patterns, reinvigorating the Malthusian imperial index of civilization. A close reading of the records of the Population Council's Ad Hoc Committee on Policy clarifies the hegemonic commitments that animated the council's activities in promoting scientific knowledge of the population crisis and fertility control as its solution. It demonstrates that the council's practices reflected a negotiation of gendered nationalist politics that pitted the council against birth control advocates and anticolonial nationalists as it sought to align the social and political elites across the globe with the neo-Malthusian pursuit of population control.

The Office of Population Research

The OPR occupied modest offices in a house not far from the Princeton University campus. Its research enterprise initially relied on a slender staff of graduate student fellows and demographic wives.[13] In this unassuming space, the OPR became a central site of midcentury demographic research in the United States. It served as a clearinghouse of population knowledge through its publication of the *Population Index,* which monthly provided a bibliography of population research, along with disciplinary news and statistical analyses.[14] The *Index* was the primary demographic journal published in the United States until the founding of *Demography* as the official journal of the PAA in the mid-1960s. Irene Taeuber and Louise Kiser, wife of Clyde Kiser, were responsible for compiling the bibliography and writing feature articles as well as the actual production process. Norman Ryder recalls that one of the obligations of fellows was to assist with proofreading.[15] Kingsley Davis, an irascible contrarian, was a prolific scholar who came to Princeton in 1944 as a research associate at the OPR and as a Princeton sociology professor. He would move to Columbia University in 1948, but the research he completed while at Princeton established him as a central figure in the field. He would produce some of the most important sociological theory regarding fertility and trained many important second-generation demographers.[16]

During the early years, the OPR also hosted PAA annual meetings, bringing together national and international leaders in the field. Participants later remembered fondly the social intimacy and intellectual vigor of these early meetings, "when everyone knew everyone else."[17] The OPR also played a significant role in the discipline's development through a graduate fellowship program, which was initially supported by the MMF and later by the Population Council. But most important, the OPR oversaw important demographic research projects, including the famous Indianapolis Study, which was the first of several multiyear investigations of the social and psychological determinants of fertility in the United States.[18] In 1937, when portions of the League of Nations administrative offices relocated to Princeton from Europe, the league contracted with the OPR to complete four monographs on population trends in Europe. At the end of the war, the Division of Geography and Cartography of the US State Department contracted with the OPR to extend its statistical expertise to the Global South. The studies made under the joint auspices of the league, the State Department, and the OPR "were integral parts of a long-run comprehensive research program in comparative international demography."[19] Particularly important were Kingsley Davis's 1951 monograph, *The Population of India and Pakistan,* and Irene Taeuber's analysis of Japanese population patterns, compiled in a 1958 monograph, *The Population of Japan.*[20] Davis's analysis, published in a series of articles in the 1940s, was the first to suggest that the potential for rapid growth in India was far greater than Western demographers had yet anticipated.[21] This potential resulted from the substantial decline in mortality and persistent high fertility. Taeuber's work, which also appeared in numerous articles prior to the book, demonstrated that Japan's demographic transition—the only one in a non-Western, nonwhite nation—was produced by a conscious program of fertility control.[22]

The OPR's involvement with international population statistics extended to the newly established United Nations through Frank Notestein's appointment as the first director of its population division. The successor to the League of Nations, the United Nations continued many of the league's administrative tasks. Although there was also a separate statistical division, and despite some tensions between the two, the population division played a significant role in the development of international demographic standards. As its director, Notestein pushed for the development of measurement standards for the variables needed to investigate "the interrelationships of economic and social conditions and population trends."[23] The population division also directed technical support efforts for the first postwar censuses in new nations of the Global South in 1950 and 1951. Those censuses provided the data for the first comprehensive population projections of the postwar period, the numbers that enunciated the population explosion.[24]

Demographic transition theory was produced by those associated with the OPR. Frank Notestein is generally recognized as the principal author of the theory. According to disciplinary histories, Warren Thompson published a prototypic version of demographic transition theory in 1929, but this version is said to have been ignored until Notestein revived the idea at the end of World War II. Thompson's early articulation may well have been pushed aside after the war because it combined a strong critique of imperial mindedness, a hearty endorsement of contraception, and a boldly pejorative perspective on "the white man's burden." His rhetoric certainly did not comport with the technocratic tone on which demographers staked their creditability.[25] As noted in chapter 4, Notestein spoke of population changes in mortality and fertility as transitions in the late 1930s. But the concept of the demographic transition awaited the wartime analysis of population changes in Europe and the Global South. Both Dudley Kirk and Kingsley Davis, who also published early articulations of the demographic transition, said that the concept was very much in the air at the OPR by the mid-1940s.[26] It also circulated widely in scholarly circles. Beginning in 1947, Frank Boudreau and Clyde Kiser organized a series of meetings to investigate population problems in the Global South. The proceedings of these meetings, which were published in the *Milbank Memorial Fund Quarterly* in 1947, 1949, and 1953, contributed to the growing body of scholarship supporting both demographic transition theory and the conclusion that "undeveloped" countries were growing "too rapidly."[27] Also in those years, Notestein, Davis, and others published multiple additional articles on transition theory in such disparate journals as *Eugenics Quarterly*, *Science*, and the *Annals of the American Academy of Political and Social Science*.[28]

Demographic transition theory picked up where earlier debates about population had ended. In the interwar years, population discourse clustered around the question of whether Malthusian principles still held. Massive increases in Western living standards and population sizes that had accompanied industrialization seemed to invalidate Malthusian mathematics. Alexander Carr-Saunders's 1922 work, *The Population Problem*, updated in 1936, provided an exhaustive summary of the current state of knowledge about world population patterns. He argued that an optimum population size existed for every society, which was based on its developmental level and resource base. Moreover, he argued, societies rarely achieved the Malthusian maximum—China and India were exceptions. With regard to the social factors shaping fertility, he identified marital practices, Malthus's preventive checks, and the deliberate limitation of fertility. He was one of the first population scientists to acknowledge contraception as a cause of the low fertility of modern nations. Yet he also recommended that the more advanced nations and more successful classes recommit to family life and reproduction as their moral duty to kin and country.[29] For others, such as Warren Thompson, overpopulation was already a pressing international problem, and rising population pressure threatened to provoke political disruption and possibly war.[30] There was substantial disagreement about solutions, the major focus of which was whether migration would work to reduce population pressure in overcrowded areas. Warren Thompson and Radhakamal Mukerjee, a prominent Indian sociologist active in international population circles, supported opening underpopulated areas to migration. But theirs was a minority position in Western nations that had just secured restrictions on immigration.[31]

There was general agreement on five key points about population within these debates. First, growth of the Western population since 1700 was declared to be unique in human history. Population scientists concluded that throughout much of human history population had been more or less stationary, balancing high birth rates and high death rates, and thus grew slowly (figure 5.1).[32] Second, current vital processes reflected the impact of past vital processes and would reverberate into the future. That is, patterns of birth and death in the past shaped the number and rate of births and deaths in the present. Fewer people born yesterday meant that there were fewer people at risk of dying today. Fewer infant deaths today would mean that there would be more women at risk of pregnancy in 15 to 20 years. Third, population patterns were the products of the "entire bio-social situation." Thus, Kingsley Davis declared, "the so-called characteristics of the population . . . such as sex and age are primarily biological, but through their variation from one group or region to another they reflect social realities."[33] Fourth, social change, and with it population patterns, followed a process of evolution from primitive to advanced, from simple to complex, from agricultural to industrial, from dispersed/rural to dense/urban settlements. And fifth, social evolution occurred through processes of diffusion of economic, technical, and social advances from centers of innovation out and down to less advanced groups, nations, races, and classes.[34] These principles meant, as Alexan-

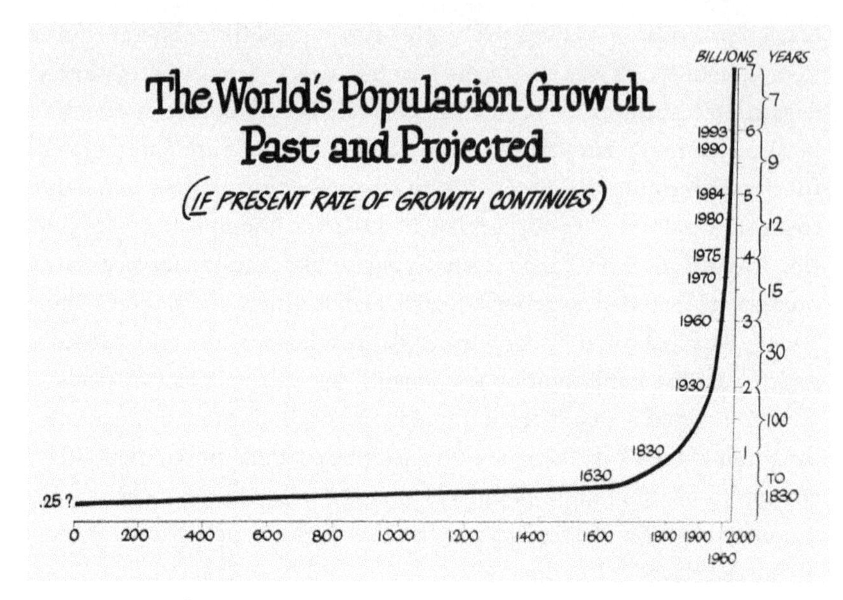

FIGURE 5.1. A midcentury US State Department representation of the "unprecedented" growth of world population. From Piotrow 1973, 4.

der Carr-Saunders noted, that "the population situation today" is "the consequence of the whole story of social evolution." Moreover, as Kingsley Davis noted of India in 1951, "A description of the population is thus at the same time a description of the social order."[35]

The Writings of Frank Notestein and Kingsley Davis

Along with the other US social sciences, demography responded to Truman's call for a "fair deal" for the emerging nations of the Global South by configuring discourses and practices that became central to the modernization regime. US foreign aid to those regions emphasized the rationalization of agriculture to improve food production through modern methods of plowing, irrigation, and pest control and development of indigenous industry, stressing resource extraction.[36] Questions about food production and natural resources always lead back to Malthus. Just as the threat of food shortages in the immediate aftermath of World War I provoked talk of overpopulation, so too post–World War II discussions of reestablishing agricultural production and trade stoked questions about the size, composition, and distribution of the world's population. Several of the initial articulations of demographic transition theory by Notestein were presented at and published in international agricultural forums.[37]

Political realignments embodied in the modernization paradigm involved important adjustments in gendered coloniality as well as creating a new economic order. Already in dispute in the late 1930s, biological theories of race differences were widely discredited in postwar social science discourse. In its place, social science located racial difference in the nexus of culture and personality. Differences were still racialized, but they resulted from culturally specific structures of family and community life.[38] Social organizations and cultural values were said to structure the rewards and incentives that shaped human personalities and activity. Kinship systems and religion became the motivational backdrop in explanations of the personality types and behavioral patterns of individuals and groups.[39] Cultural differentialist social theory still fulfilled the commitment to colonial difference as the underlying cause of the "fate of societies and peoples."[40] In terms of the modernization paradigm, the new discourse of cultural difference highlighted indigenous gender-based social and economic structures as the distinctive features differentiating "traditional," "Third World" cultures from the "advanced" cultures of "the West."[41] In this way, modernization theory still relied on an evolutionary logic of social change with sexual reproduction at its core. It refurbished longstanding imperial narratives of cultural competencies that marked the distinction of bourgeois Western culture from its others.[42] This was certainly the case with demographic transition theory, which exported the "as-if" woman of natural fertility who, coupled with the "backward-looking" "traditional" Malthusian individual, engaged in the "functionless reproduction" that produced the crisis of rapid population growth.[43] Conflating the individual and the aggregate, population figures and charts indexed the "backwardness" of rural peasants of Asia and signaled the dire straits emerging nations would find themselves in if they did not control population growth. New demographic data on growth suggested that intervention would be necessary to normalize the population patterns of "less developed countries." That normalization required the acquisition of the small-family system through family planning, which demographers championed as the Western achievement that would ensure successful economic development.

The small-family system coupled demography's differentialist narrative of modernization with functionalist sociology's sex-role theory in a renewed formation of hegemonic masculinity. Technology accounted for economic change, but family—the social institutions and structures in which gender relations were said to matter most—depended on customs and traditions, "the heritage of past ages," which entailed "man's deepest beliefs."[44] Within the functionalist sociology, the family began as the primary unit of organization within primitive societies. It was the location of both productive and reproductive activities. As societies evolved into complex agrarian systems, large extended forms of family shaped the lives and personalities of individuals, fitting them to the roles they were destined to fill. Such kinship systems were said to exemplify timeless, classical patriarchy.

With the rise of modern industrial societies, the social role of the family was said to have diminished. Socialization of children became the main task of family life, as urbanization and industrialization moved the organization and activities of work, education, and health care to factories and the state. As elaborated by Kingsley Davis, who was a protégé of Talcott Parsons, functionalist sociology defined the normal American family as one in which men dominated as instrumental economic actors, while educated but still economically dependent women did the expressive work of caring for their children. As Davis argued in his 1948 textbook, *Human Society,* the family environment built on these roles, and the context of urbanization provided the grounds for the development of modern character traits and the cultural competencies needed to sustain economic progress individually and nationally.[45] Those traits, which Davis and the broader field of sociology identified, included individualism, acquisitiveness, an achievement orientation, and a focus on the future, all of which behavioral scientists linked to the inclination toward democratic practice and the penchant for economic success.[46] Demographers such as Davis added rational reproductive restraint to this mix. At home, social stability, well-adjusted children, and good citizenship depended on stable life in nuclear families. Abroad, modernization projects should cultivate the "nexus of cultural traits valued as 'progress'"—the "empirical outlook on life," individualism, achievement, and family limitation.[47] The smooth functioning of families still depended on conventional gender differences, however, reinforcing the midcentury consensus that after marriage women belonged in and to the domestic sphere.[48] The enlightened men of modern societies gave women greater freedom, within limits. Thus, women were no longer needed for endless childbearing, but they remained primarily responsible for children and family life.

On the surface, the demographic transition described by Notestein theorized a precise and automatic relationship between population change and economic development (figure 5.2). "There is no mystery about the rapid population growth that accompanies modernization in undeveloped areas."[49] Notestein's descriptions began with the archaic societies described by Malthus: traditional societies in Europe's past and among "today's backward people" that were said to exhibit a pattern of high mortality and high fertility rates and, therefore, slow if any growth.[50] High mortality naturally prevailed in them because "societies with low levels of technical skill are inevitably poor, ill-housed, ill-clothed, ill-fed and subject to the uncontrolled ravages of disease."[51] Moreover, given the natural excess of human fertility, Malthusian logic dictated that mortality had been high for most of human history. Otherwise, the population would have grown so much that the planet would already be overrun by multitudes of human beings. The example often cited is that a mere 1% growth rate for 2,000 years of "the Christian Era" would yield a population of "60–80 billion."[52] Given high mortality, it follows that "populations must have high birth rates to match their inevitably high death rates. Those that did not have high birth rates are no longer represented in the world. The very existence of such populations in the face of the toll of heavy mortality proves that the birth rates are high, and that the societies have developed the social structures essential to produce and maintain the birth rates."[53] Notestein enumerated an impressive list of the social structures supporting high fertility in premodern societies, saying, "All such societies are therefore ingeniously arranged to obtain the required births. Their religious doctrines, moral codes, laws, education, community customs, marriage habits and family organizations are all focused toward maintaining high fertility."[54]

In contrast, modern, urban, industrial societies exhibit low mortality and low fertility rates, and grow slowly.[55] The social progress and prosperity produced by modernization triggers an automatic shift from high to low rates, which constitutes the transition for which the

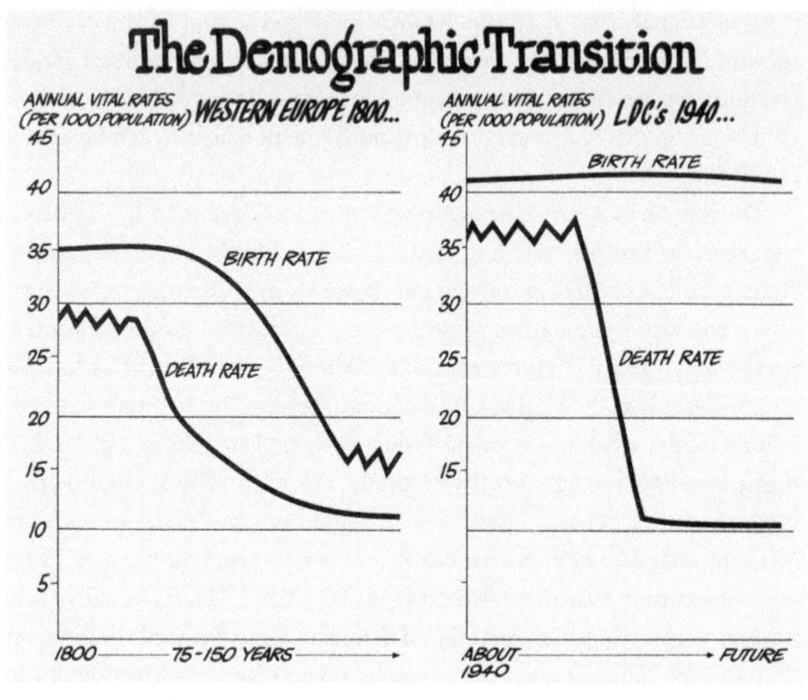

FIGURE 5.2. A midcentury US State Department representation of the demographic transition. From Piotrow 1973, 6.

model is named.[56] Europe and its settler colonies provide the standard case. In "the middle of the seventeenth century," "an era of peace and domestic order" led "shortly afterward" to "agricultural innovations" and, more importantly, "industrial innovations." "In short, the whole process of modernization in Europe and Europe overseas brought rising levels of living, new controls over disease, and reduced mortality."[57] Declining mortality rates initiate the transitional phase of rapid population growth.[58] Notestein and his colleagues concluded that mortality declined first because it responds most immediately and directly to a rising standard of living and improved public health. The dramatic decline in mortality in Sri Lanka (then Ceylon) after DDT-based mosquito eradication efforts in 1945 and 1946 served as the prime example of this aggregate effect. This decline also heralded what Kingsley Davis identified as unprecedented and far greater than predicted declines in mortality rates in colonial regions of Asia.[59]

Public health measures, however, were said not to alter the social organization of individual lives. Such changes, Notestein argued, "influence only the externals of life and leave the opportunities, hopes, fears, beliefs, customs and social organization of the masses of people relatively untouched." These were "the factors that control fertility."[60] "The reduction of fertility . . . awaits a change in age-old social organization and in the motives governing reproduction"[61] because a society's fertility pattern "depends on the social organization, customs, and beliefs from which arises the aspiration of its people with respect to family size. These matters, the heritage of past ages, lie at the core of the society and are scarcely modified by relatively small changes in government, in modes of production, and in sanitation."[62] Therefore, fertility "responds scarcely at all" to the "initial" changes brought by economic and political progress. "Even under the impact of a rapidly shifting environment," the values and beliefs about family size "change only gradually."[63] As Kingsley Davis argued, in premodern societies fertility was controlled "not" through "deliberate control" but indirectly, through "institutional practices. . . . If controls [were] indirect and institutional, a long time [would be] required for them to be abandoned and deliberate controls adopted."[64]

Fertility, said to result from the nondeliberate actions by individuals whose family life was mired in timeless customs, religions, and community mores, remained high in the early part of the transition phase. Gradually, however, as time passed, social changes altered the structures of individual rewards and incentives, and fertility would begin to fall as well. That is, "urban-industrial development yields much more than an enlarged economic product. It is also an essential means of stimulating the social changes from which the small family ideal emerges." Notestein cited the declining productive value and higher cost of raising children as the primary example of changed reproductive incentives.[65] Davis noted, "It is only after the successful preservation of life has resulted in larger families, and these larger families have proved an embarrassment to the individual in the highly urbanized and mobile structure of modern society, that he seeks a way around the full practice of his high fertility mores."[66] When the decline in fertility was sufficient to bring about a new balance of low birth and death rates, the transition was said to be complete. The modern balance of low mortality and fertility rates represented "a much less wasteful kind," one that allowed a stationary population to reemerge and normal slow growth to resume.[67] In the midst of the transition, however, the "lag in the decline of fertility behind that of mortality" could produce substantial population growth.[68]

In the West, the transition took three centuries, during which the population of "European extraction" increased nearly "sevenfold."[69] Demographic data worked up by OPR demographers during and just after the war suggested that the "potentiality" for growth in Asia and the Middle East was significantly greater than had been seen in the European transition.[70] The alarming specter of "rapid and gigantic expansion" loomed over these regions because "there [were] already teeming millions in Asia,"[71] and their larger populations meant that a moderate growth rate would yield "a huge absolute increment."[72] This demographic threat occurred because mortality was said to have declined more substantially and more quickly in colonial and former colonial regions than was anticipated by their level of development. Thus, the population dynamics of emerging nations deviated from the standard case exemplified by Europe and its settler colonies. Their mortality rates started to drop earlier and fell faster than had occurred automatically in Europe. This "peculiar" situation occurred because of "partial diffusion" occasioned by colonialism.[73] These nations had begun to modernize, but as Davis noted of India, "modernization has not resulted from an internal impetus to which all phases of her civilization have contributed, but has been borrowed from the outside." In colonial areas, "modernization has proceeded mainly in those matters that are profitable, easy, or sentimentally desired by the Europeans"—matters that were "economically and militarily advantageous."[74]

In a narrative that transfigures the violence of colonialism into benefits, Notestein and Davis contended that colonial administrations had produced domestic order and improved transportation, sanitation, and communication to sustain their own population and economic activity.[75] Moreover, Davis noted that "the control of mortality appealed strongly to Western humanitarianism."[76] Thus emerging nations had benefited from the death-reducing technologies of Europe.[77] And, as with public health measures in the West, these changes lowered mortality but did not change the essentials of the social life, customs, or mores of the masses. Thus the high fertility rates said to be traditional in these areas continued unabated, precipitating gigantic and accelerating population growth.

Western conceit also led demographers to conclude that the humanitarian motives of the West probably exacerbated the normal lag in fertility decline in the Global South because, they argued, in "enlightened colonial regimes there had been considerable protection of native customs, and religions, and social organization all of which foster the maintenance of high fertility." The social structures of traditional societies, the cultural practices that organized normative gender relations, including "family form and function, religious doctrine, and community custom," were represented as timeless, sentimental, and untouched by colonialism. Notestein declared that the crux of the population problem was that "technologically advanced nations have disseminated and imposed that part of their culture which reduces mortality, while withholding, or at least failing to foster the transfer of, that part of their culture out of which the rational control of fertility and the small family pattern develop. The population grows a good deal as it did in the West, but unlike the situation in the West, the growth stage has not been accompanied by the social changes that eventually lead to an end of expansion."[78]

He concluded that the coupling of reduced death rates and the high fertility said to be typical of traditional peasant societies produced growth seldom seen in world history. Thus, consistent with development discourse, the narrative of population change in the Global South constructed those regions as abnormal and in need of intervention to correct their unbalanced demographic situation.[79] In the face of rapidly declining death rates produced by borrowed techniques, without fundamental social change, high birth rates, even if beginning to decline, would inevitably result in "too rapid" growth. It was simply a matter of mathematics. If the mortality rate was 30 per 1,000 and the fertility was 40 per 1,000, there would be an excess of 10 per 1,000, or 1% growth per year. If, however, the mortality rate dropped to 20 per 1,000 and the fertility rate remained the same, then the annual growth rate would double to 2% per year. Moreover, demographers hastened to point out that even if fertility rates dropped immediately, the rapid growth would reverberate far into the future. This is part of the dynamism of population. Like the chain reaction of a nuclear bomb, population growth breeds its own continuing expansion merely because in each successive generation there are more (surviving) women of childbearing age than in the last.[80] Even with declining birth rates, rapid growth would occur unless the decline were sufficient to cancel out the number of additional childbearing women.

Such a drastic change seemed very unlikely without conscious effort. In fact, demographers concluded, there had not yet been any "appreciable" fertility decline in the Global South. But they postulated that mortality would continue to decline, more rapidly than fertility. Moreover, new nations of the Global South would start the transition with a lower level of economic development, a larger population base, and estimated fertility rates that were higher than those said to have prevailed in the European transition. But, again, reliable data continued to be scarce. And unlike the European transition, the outlet of "unpopulated" areas for migration no longer existed.[81] The mathematics alone portended a dangerous demographic situation in much of Asia and stoked panic about population growth.

The mathematics of population increase was not the sole cause of concern, however. The Malthusian logic of population and resources also spurred alarm. Notestein and Davis doubted that the automatic processes of the demographic transition would occur in the Global South because population increases would undercut any rise in the standard of living and "retard" industrial and economic development.[82] As discussed in chapter 6, precise estimates of the impact of population growth on economic development were made in demographic publications and Population Council deliberations. A 1% rate of population growth was said to "require investment of 4% or 5% in capital formation to keep constant per capita income."[83] Moreover, demographers reasoned that high fertility and declining infant and childhood mortality produced disproportionately young populations. Such populations would undercut development efforts because the smaller proportion of the population in the economically productive ages would need to support the larger proportion of the dependent/unproductive population. This dependency ratio would require even greater strides in per capita productivity to keep up with population growth.[84] However, "future population growth" would likely be "of a magnitude that sooner or later threaten[ed] to become a serious obstacle to political and economic development."[85] Thus, "the prospects of removing the demographic threat [did] not appear promising."[86] It was only a matter of time before efforts to raise the standard of living were swamped by the growing numbers. Together, the population and economic factors led Kingsley Davis to the "melancholy conclusion" that "population growth has become a frankenstein" in places like India.[87]

Again, as in Malthus, the sexual economy trumped political economy. Uncontrolled natural reproductive processes, the "as-if" women of natural fertility statistics, in concert with traditional culture, undermined modern economic plans and actions.[88] "The imbalances in fertility among various countries of the world" constituted the underlying cause of the differential fate of nations.[89] Each nation's mortality and fertility patterns determined its location on the transition graph. Those mortality patterns are shaped by science and technology, and fertility patterns are affected by the heritage of the ages. Affirmative control of the excessive fertility of female bodies marks the rise of the rational disposition for the modern nation, "the willingness to solve its problems rather than accept them."[90] Through population dynamics, the weight of the past, the dampening effect of tradition, dragged down the present and risked the future.

Demographers' doubt and alarm were not based in biological assumptions of racial difference, however. According to Frank Notestein, "the available evidence" suggested that people "having different racial origins and widely different cultural backgrounds modify their reproductive behavior in response to changing environments in the same way that persons of European origins have done."[91] Kingsley Davis argued as well that only the fallacy of inherent racial difference would lead one to conclude that Asian nations would not develop the same rational control of fertility as they advanced. Davis and Notestein relied on the research of their OPR colleague Irene Taeuber on Japan's demographic transition. That Japan had been able to make the change to low fertility and mortality as it modernized "proved" that biological racial differences would not prevent the demographic transition or the rational control of fertility.[92] It was Japan's culture shift that mattered more. Culture in the form of modern family and community mores encouraged the small-family system and the conscious control of the excess natural fertility of women's bodies.

Given the atypical population patterns of emerging nations, demographers concluded, intervention was required to bring their demographic patterns into proper alignment before it was too late. "The demographic account [had] to be balanced sometime."[93] They argued that "the East, unlike the West, [could not] afford to await the automatic processes of social change, incident to urbanization and industrialization, in order to complete its transition to an efficient system of population replacement."[94] The base populations were too large to be sustained by any reasonable estimate of economic progress. Without sufficient development, it was unlikely that standards of living would rise sufficiently to spur the motivation to limit fertility. If such motivation remained weak and fertility remained high, a "vicious circle" of "too rapid" growth and economic insufficiency would ensue. Since the West's imposition of death reduction technologies had caused the problem, demographers argued, the West had an obligation to provide solutions to the population problem resulting from continued high fertility. This "require[d] a complete and integrated program of modernization" in which "every device for the creation of a social setting favorable to reduced fertility" should be applied.[95] The implicit conceit here is that the United States was best positioned to do the job. The demographic transition thus fit neatly into development discourse, "the organizing premise" of which "was the belief in the role of modernization as the only force capable of destroying archaic superstitions and relations," and the US culture and economy offered the best model for that modernization.[96]

As was the case in relation to the question of contraceptive effectiveness, Notestein's first response was to call for more research into the best means for promoting lower fertility rates. In advance of such research, however, he reasoned from the example of the West that "the small family ideal, and the rational control of fertility as a means for its attainment, develop most rapidly in a society that stresses the functions, welfare, and the opportunities of the individual . . . "[97] With "urban industrial living" in the West, the individual

> came to depend less and less on the status of his family for his place among his fellows. The station to which he was born gave place to his accomplishments and possessions as the measure of his importance. Meanwhile, the family lost many of its functions to the factory, the school, and commercial enterprise. All these developments made large families a progressively difficult and expensive undertaking; expensive and difficult for a population increasingly freed from older taboos and increasingly willing to solve its problems rather than to accept them. . . .

> The social aim of perpetuating the family gave way progressively to that of promoting the health, education, and material welfare of the individual child.[98]

Societies made up of such individuals "develop a rational and materialistic outlook on life." They "view man as the master of his own destiny" and come to see "the deliberate control of fertility to be as reasonable and desirable as that of mortality." Such societies provide a "favorable milieu" for "the small family ideal" to emerge.[99] The achievement of the small-family system through competent reproductive restraint marked the disposition of modern man.

Notestein offered several suggestions for encouraging greater individualism. He did not call for direct changes in existing cultural institutions; rather, his recommendations focused on "freeing" individuals from the influence of those institutions. Thus he argued: "Although the rate of childbearing is greatly influenced by the social milieu of custom, religious belief, social organization and the like, it should be possible to modify reproductive behavior to some extent without fundamental changes in other components of culture. No social system, however coercive, maintains absolute homogeneity of behavior. All systems have their dissident extremes open to innovative suggestion; all have those who conform only because of the absence of alternatives."[100] Thus, the best course of action was to provide alternatives that would stimulate interest in fertility control among those already inclined to change. He primarily suggested expansion of educational and economic alternatives to traditional livelihoods, preferably beyond local villages. He reasoned that "city life, a milieu of technological development, education, and other means of attaining contact with the modern world should favor such changes."[101]

As the above quotations make clear, in midcentury demographic narratives, the individual who was or was not motivated to control fertility was masculine.[102] He was a neo-Malthusian man. As we saw in chapter 2, Malthus's population principle calibrated masculinity and civilization by Orientalist measures of sexual civility and marital form, thus giving quantitative grounding to the imperialist truism that the level of civilization could be judged by how a society treats its women. Demographic narratives represented gender relations in "backward nations" as timeless patriarchal systems in which women were valued only as childbearers. The heritage of the ages, said to determine fertility rates in those nations, remained untouched by the forces of modernity. In contrast, the neo-Malthusian man who animated transition theory adhered to the logic of modern society and its small-family system. In so doing, his individual rational restraint of fertility encoded (or failed to encode) modern market logic and fertility values in equilibrium. "The prerequisites of manhood"[103] require that he engage in an assessment of his ambitions, accomplishments, and responsibilities when making his reproductive decisions. He consciously controls costs by limiting his family size through both the prudent delay of marriage and effective contraceptive practice. If mentioned at all as individuals, women enter the transition model as a dependent wife trailing after him.[104] Where she is described as having any specific interests or motivations, they are either wholly synchronized with or subordinated to those of her husband and culture.[105] Demographers dismissed any suggestion that women's desire to escape "perpetual child bearing" was "synonymous with a desire for birth control."[106] To them the first was an abstract wish; the second was a specific desire. Assuming that the heritage of the ages in which she was mired was even more restrictive and disapproving of contraception on moral grounds than Western nations were, they assumed that she was always already uninterested in contraception. Instead she awaited the educational efforts of the programs demographers endorsed. Those programs focused on reducing aggregate rates, not enhancing women's rights or well-being.

This is not to say that demographers ignored the question of women's status. To the contrary, demographers recommended changes in women's social lives as one means of altering fertility patterns.[107] Thus, for instance, research conducted by Kingsley Davis in the mid-1950s suggested that raising the average age of marriage was potentially the single most effective means of reducing fertility quickly because it reduced the length of time in which uncontrolled fertility could occur. Using the extensive and readily available figures for Switzerland, he concluded that "in terms of the potential fertility lost by late marriage, the figure is approximately 64%."[108] Notestein also recommended greater education and work outside the home for this reason. However, demographers were not seeking to change women's social roles. The changes they supported were intended merely to delay marriage and childbearing. The goal was to eliminate "functionless reproduction," not patriarchal marriage. They sought only to remediate the cultural flaws of "the peasant" and "his wife" that had produced disequilibrium in global population patterns.[109]

Further, consistent with colonial discourses, changes in the relations of men and women were only required in the developing world, not at home. The problem in demographic terms was that the developing world had the wrong values for its new circumstances. Its familial customs, that is, its gender order, needed to be updated so that its fertility rates would begin to follow the "normal," that is, modern, Western, balance of low vital rates. Aligning with the Orientalist conceit about its greater gender equality, the rest of the world needed to follow the example of Western nations and give women greater (but still not equal) economic roles and political rights. In this narrative, modern Western gender relations developed naturally as a result of social progress and enlightened thinking. Changes like these were said to be nearly automatic in an urban-industrial community, a claim that elided women's ongoing political struggle for equality and autonomy. Moreover, the narrative reaffirmed the superiority of Western masculinity, as the rest of the world was represented as already behind with regard to recognizing the justice in raising the status of women to their proper place on the pedestal of bourgeois femininity.

Demographic transition theory thus provided a graphic representation of the value of the nuclear family for the modern world. Following functionalist accounts of social change, the changes in women's roles that accompanied modernization ensured the functional stability of family life in urban, industrial societies.[110] Small families of heteronormative couples were best suited to raise the next generation. Well-reared children, ready to meet the demands of modern industrial life, needed educated modern mothers who, informed by science, would maintain the optimal environment for their development. Demographic concern for women's social status was thus very limited. Women of the demographic transition were to be modern wives. The ambitions, accomplishments, and status of the individual were still his, not hers. She still derived her status from successful marriage and motherhood within the modern nuclear family.

Similarly, in demographic measures women became objects of analysis. In the single-sex model of stable population, women's bodies were the event-horizon by which demographers measured fertility rates. Their bodies became the object through which to track the couple's reproductive decisions and achievement of normative family life. But this focus on the individual woman should not be mistaken for concern for her individual well-being or empowerment. As in the case of the contraceptive effectiveness studies discussed in chapter 4, such measurements lack any conceptual space in which to observe or theorize gender-specific sexual or reproductive interests, motives, or constraints of individuals.[111] Women's bodily processes thus indexed the modernity of men and nations.

In another instance of slippage between the individual and the aggregate, the figure of the excessively fertile village woman, the peasant's wife, personifies nations and regions. For instance, Dudley Kirk predicted that "Asia may never be successful in extricating herself from the vicious circle of poverty, ignorance, and over-population that has been her lot in recent centuries."[112] Likewise, the two nations demographers identified as high-fertility nations, China and India, were also personified as feminine. So, for instance, in discussing the situation in China in 1944, Dudley Kirk asserted that the trend of population growth and economic development there would "depend upon how quickly she absorbs the pattern of birth control."[113] Kingsley Davis also generally referred to India as "she." For example, in the introductory chapter to his 1951 book, he justified the research by saying, "Certainly she has had more usable data than any country of equal backwardness and this fact constitutes one of the strongest reasons for making a special study of her population problems. She exemplifies some of the major difficulties besetting the heavily peopled rural economics of the world."[114] Demographic transition theory thus both forecloses close scrutiny of gender relations as they are lived in the individual beneath a differentialist tale of social progress and genders traditionalism as feminine.[115] The feminine figure of excess fertility is an object of threat—the fuel of the population explosion—and control of her was necessary to rescue the world from a grim future. Moreover, it positions the demographic knowing subject as the authority on the risks she posed and the value of efforts to contain those risks.

Notestein and Davis concurred that rapid industrialization was the main means by which to foster modern individualist dispositions in former colonial regions and thereby check too-rapid growth. Davis differed from Notestein in drawing the more "melancholy" conclusion that such efforts were not likely to be successful.[116] Reiterating the pessimism for which Malthus was so famous, Davis doubted that emerging nations would be able to achieve the rapid and steep decline in birth rates needed to prevent population growth from overwhelming economic development efforts. Notestein initially agreed that without measures to contain fertility at dramatically lower levels, "a sober consideration of the existing situation leads one to expect that catastrophes will in fact check rapid growth."[117] Although both continued to disparage the birth control movement itself, they embraced the deliberate control of fertility through contraception. Yet, here again, both were pessimistic about the effectiveness of efforts to promote contraceptives.[118] This pessimism follows logically from the gendered coloniality of their conceptualization of social evolution and the natural excess of fertility.

The weight of Notestein's conceptualization of fertility control is apparent in the earliest studies of population dynamics in Asia, such as the 1948 survey of population in the Far East funded by the Rockefeller Foundation. Along with Notestein, the group included Irene Taeuber and Marshall Balfour. The report, published in 1950, canvassed Japan, Taiwan, Korea, China, Indonesia, and the Philippines. The overall assessment was pessimistic. In particular, the report concluded that no existing contraceptive method was suited to meet the need in rural Asia, which it described as societies of "poor and retarded populations." That is, the report argued, "the problem of reducing the fertility of peasant populations has two parts, one relating to motives and the other to means. Without the motives for family limitation no means will suffice; and with weak motives efficient means are required. . . . At present, there are no means of control which meet the needs of Asia's rural population."[119]

Apparently, none of the folkway methods that Notestein evaluated in *Controlled Fertility* as reasonably effective in producing the Western transition could meet the need for population control in the Far East. As he concluded in his initial articulation of the demographic transition, the "dissemination of contraceptive knowledge" was "of little importance" because "indirect evidence" demonstrates that populations in high-fertility societies were not interested in contraception and did not "make effective use of methods" that were "normally at their disposal" (withdrawal).[120] That is, reiterating the logic of *Controlled Fertility,* high fertility rates indicated that folk methods which were, he assumed, widely known and reasonably effective, were not widely used. Fertility was high because motivation to limit it was weak. Therefore, existing methods would not be sufficient to meet the need for an immediate and substantial reduction of fertility rates. Need here is predicated on the assumption that weak motives required highly efficient means to effect aggregate change.[121] Again, Notestein's discussion completely overlooked the potential for differing perspectives on motives and means between husbands and wives that would complicate the question of contraceptive use.[122] Kingsley Davis largely concurred with Notestein. In a review of the situation published in the *Eugenics Quarterly* in 1955, he concluded, "All told, the big agrarian societies appear to be the ones where the institutional supports to functionless reproduction are likely to be the strongest, to persist longest, and to cause the greatest demographic imbalances. It is of great sociological interest to study the process of change by which this functionless fertility is gradually eliminated through alterations in the reproductive institutions."[123]

Based on his assumptions about means and motives, Notestein did not initially support the provision of contraceptives in nations of the Global South. Such efforts would be futile until a means that would be effective in the face of weak motives was developed. Moreover, such efforts might stir opposition to family limitation in populations presumed to be more conservative than those in the West. Again, Davis largely concurred and remained a critic of the notion that fertility control programs would in and of themselves succeed in reducing fertility sufficiently to prevent disaster.[124]

The combination of declining mortality rates and high fertility rates produced in population projections based on the 1951 All India Census moved Notestein to take up a more interventionist perspective.[125] Consistent with his earlier views on the role of motivation, Notestein argued that the "first step" was "to provide adequate services" to the "already motivated."[126] By the time he took charge at the Population Council in 1959, Notestein had become a major proponent of family planning as the means to produce the demographic transition, especially in Asia. By that time, the research was "pretty well advanced" and the "dangers of getting into action [were] far less than they were." Moreover, "a clear reduction of ideological and religious controversy" had occurred, easing "the problems of international technical assistance and of national program formulations." Thus, he was optimistic that "the world's extremely dangerous problems of rapid population growth can be greatly reduced in the coming decades."[127] By this time, Notestein had already been involved in action programs, having traveled to India on behalf of the council on several occasions to confer with population scientists and government officials.[128] When Notestein and the Population Council became directly involved in contraceptive research and services, withdrawal was not among the methods it encouraged. Under Notestein's leadership, the Population Council invested heavily in the development and distribution of intrauterine devices (IUDs), a contraceptive designed for situations of weak motivation and "high need."[129]

The Population Council

Contraceptive research pursued by the Population Council under Notestein's leadership merely extended the council's existing interventionist approach to include direct clinical practices. From its founding, the council linked demographers, funders, and government officials in efforts to develop both the communications strategy for promoting population control policies and the policies themselves. An initial conference to discuss the possibility of establishing the council was held in June 1952.[130] This meeting occurred at a historical juncture at which efforts to begin practical efforts to address the population crisis through the United Nations were stymied by Cold War politics and Vatican opposition. Birth control advocates were nonetheless working toward a new international family planning organization, which would become the International Planned Parenthood Federation in 1952.[131] In this context, Frederick Osborn renewed his efforts to establish a scientifically based population advocacy organization. He collaborated with his friend John D. Rockefeller III to set up the Population Council with the stated goal of supporting the pursuit and dissemination of the science of population.

As previously noted, Frederick Osborn disdained birth control activism, calling it propaganda. He favored a quiet approach, using science to pave the way.[132] Opposition from the Vatican, its allies at the United Nations, and Catholic organizations in the United States convinced him that the issue of contraception was still too controversial to pursue openly. A Eurocentric frame of reference led him to expect even greater resistance from religious bodies in the Global South and to discount local birth control organizations in the region. After confrontations at the 1954 UN-IUSSP jointly sponsored population conference, he also became concerned about the racial and anticolonial "sensitivities" stoked by "communists."[133] Given Catholic opposition coupled with what he saw as the continued disruptive activities of birth control advocates, the first step in his mind was to address the "sensitive problems of public relations" in promoting the spread of population control.[134]

Yet, as Phyllis Piotrow notes, despite its stated commitment to pursue science and eschew action, "the Population Council was drawn very quickly into a more active role," because, she says, "its professionals were among the best qualified to advise developing countries on what action to take." Whether the best positioned to provide "advice," the social network built by the council ensured that they were often the best known and the best funded. As the core institution of demography's international research infrastructure through the early 1970s, the council "bureaucratically administered and controlled" the vast majority of population research and family planning programs in the Global South.[135] It sought to provide a "scientific base in as many countries as possible on which [could] be built the action programs of the future."[136] At the beginning, however, this outcome was not guaranteed.

In order to determine the best approach for the council, Frederick Osborn established the Ad Hoc Committee on Policy, consisting of council board members, staff, and invited experts. The committee met eight times between 1955 and 1957 to consider the "philosophy and public relations of the rational control of family size." The purpose of the committee was to define the "major problems of population in different parts of the world," indicate "steps to be taken in their solution," and propose "means of meeting sensitive problems of public relations involved in such solutions."[137] Frederick Osborn and the council's professional staff, including Dudley Kirk, organized the committee meeting agenda, invited experts, and prepared staff reports.[138] Frank Notestein was a regular participant both as a board member and as an expert rapporteur. Frank Lorimer, who served as secretary of the IUSSP, was also a regular participant and rapporteur. The records of the committee inscribe the constellation of issues with which the council wrestled as it worked to situate itself as the leading private organization on population issues and define its approach to the problem of "too rapid growth"[139] in regions that were already "overpopulated."[140] Frederick Osborn compiled a summary of the committee's proceedings in *Population: An International Dilemma,* which was published by the council in 1958.[141] The book's chapters correspond to the topics of the committee meetings, first summarizing the problem demographically and then examining the factors affecting birth rates and assessing the best approach for Western nations and organizations to take to address the problem.[142] Its narrative hangs on the logic of the demographic transition and names industrialization as the trigger for which Asia could not afford to wait.[143]

The dilemma confronting the council—how best to influence foreign nations and their citizens to alter fertility practices in order to prevent dangerous amounts of population growth—derived from the following Malthusian "chain of logic" articulated by Notestein and Davis. Clearly, "too rapid population growth" was a "handicap to rapid economic and social development." But birth rates are the result of individuals acting on personal considerations, not on "national considerations." So a transition from high to low fertility required "stimulat[ing]" "latent" personal motivations. And, shaped by culture and "custom morality,"[144] individual "motivation is created indirectly" by "modernization (i.e. socio-economic changes attending urbanization and industrialization, public education, enhanced status of women, and higher standards of child care)."[145] However, aside from the "delicacy of interference" in the "most intimate domestic affairs" that advocacy of family limitation involved, there was also the "hazard of rousing intense anti-colonial and racial consciousness," which could be exploited by America's Cold War enemies—the Communist nations and factions in emerging nations. Because as Dudley Kirk notes, "what could be more dangerous material for Anti-American propaganda than the idea that rich, white Americans want to restrict the growth of colored Asian and African peoples . . ." The Western motivation would be interpreted as stemming from a "fear" of "the rising tide of color." In a footnote, Kirk observed that "the important point is of course not whether the assertions are true, but rather the fact that they are apparently widely believed."[146] Moreover, even in the West, disagreements existed about when and whether fertility control should be a matter of public policy. Thus, the committee's task was to determine what the best approach would be on the part of the US government, private organizations, and private individuals to influence national population policies and family limitation practices in foreign nations without stirring up resentment or resistance.

The committee membership encompassed a range of opinion on fertility control, including members that council staff hoped to persuade. On the one hand, William Gibbons, a liberal Catholic priest and sociologist then at Loyola College, was a member. Gibbons and Philip Ashby, a Princeton professor of comparative religion, contributed papers on religion and family limitation at the third committee meeting.[147] On the other hand, Frances Hand Ferguson, former president of the PPFA and the future vice president of the IPPF, participated as a member.[148] Indeed, she was the only woman member of the committee and one of only two women to make presentations at a committee meeting.[149] The records demonstrate that the council staff and leaders showed deference to the position of "our Catholic friends."[150] However, council members took a judgmental stance toward birth control organizations.

Seeking to blunt the church's objections to fertility control, the committee discussions enunciated two concepts: responsible parenthood and custom morality. Osborn asserted that responsible parenthood was a common value held by all members of the committee; the differences dealt with the means to achieve it. Custom morality referred to the customs by which local communities managed their domestic relations, whatever larger religion(s) they followed. With this concept, the council leaders hoped to establish a rapprochement with liberal religious leaders on family limitation. Short of that goal, the concept allowed the council to use the gap between official religious doctrine and local practices to support family limitation around the world and still express deference to Catholic doctrinal opposition to contraception, which, Gibbons made clear, would not change.[151]

Conversely, repeated allusions to publicity and propaganda in the records refer to an additional question the committee assigned itself, which was to determine if "the whole direction of the birth control movement" represented a "disruptive" approach.[152] Early staff reports argued for maintaining distance from Western-inspired family planning programs, which they assumed all existing programs were, saying that they were "lacking a well-informed sense of direction." Instead of helping, such efforts might "actually retard the change to more rational solutions."[153] Not surprisingly, given deference to religious leaders, its analysis of the methods of influence the topic required, and its ongoing contest with IPPF for authority during these years, the committee concluded that the international birth control movement's efforts were disruptive and that the council would do best to steer clear of them.[154]

After the initial meetings, which reviewed the political, cultural, and moral implications of family limitation policies, the committee concluded that the US government should not take any specific stand on population issues. Instead, it should pursue "a neutral, but not negative policy."[155] It should not promote family limitation, but it should willingly support such programs when asked for assistance by other nations. This language was picked up by the Draper Committee Report, which in 1959 recommended that the US government should respond to requests for family planning assistance as part of its foreign aid programs.[156]

In addition, the committee concluded that private organizations, if discrete, could work indirectly to stimulate indigenous concern by providing aid for "better censuses, better training, and better studies."[157] Through personal contacts and better data, the committee members contended, the reality of the situation would become apparent to the "informed public" as well as to local intellectual and political leaders.[158] The reality that was apparent (if unproved) was that "too rapid" population growth would inevitably undercut economic development efforts. Once clear to them, this reality would persuade national leaders and individuals who sought a better standard of living to limit fertility.[159]

While the council cheerfully asserted that it rejected the strict Malthusian numbers and resources argument, its perspective nonetheless reaffirmed both the argument and the associated tenets of Malthusian differentialist masculinity. Moreover, it conflated Malthusian individuals and manly states.[160] That is, the European experience of population change was said to derive from a "social structure amenable to entrepreneurial activity" and "an approach to personal relations between men and women that may have influenced later trends in the control of fertility." That approach involved individualism and the male-headed nuclear family. Similarly, Osborn made clear that the decision to marry and have children is man's because he is responsible for the care and well-being of his wife and children.[161] If only others would follow the logic of neo-Malthusian family limitation, the Western bourgeois standard of living would become a realistic possibility for both citizens and nations of the Global South. Without restraints on fertility, these emerging nations would continue the "vicious circle" of high fertility, poverty, and misery.[162]

The final two meetings of the committee added the question of the population situation in the United States to the question of the best approach to advocate population control "overseas."[163] By including the United States, the committee sought to respond to "Communist propaganda" claiming that Americans were only concerned with the fertility of others.[164] However, as the record of those discussions makes clear, the prospects of rapid population growth at home did not stir the same anxieties about economic growth among committee members, despite the sharp increase in US birth rates after World War II. The participants assumed that, unlike the situation they described in India, the American economy was sufficiently developed and robust enough to absorb the additional population, at least until later in the century.[165] Nonetheless, following Frank Notestein's observations, the committee concluded that population growth would have to come to an end in the United States at some point.[166] While economic circumstances did not warrant immediate implementation of family limitation in the United States, the old eugenic concern about fertility differentials between social groups did. Continued higher fertility among the "socially handicapped" warranted family planning programs because of the negative impact it would have on population quality in terms of social and genetic inheritance.[167] But Osborn's publication of the committee's recommendations was careful to endorse voluntary limitations in the new language of "equal opportunity," saying "all classes" in the United States should have equal access to information and means of family limitation.[168]

In other nations, although family limitation was urgently needed, the committee concluded that US groups could not effectively engage in a mass campaign to persuade individuals to practice family limitation or provide clinical services. On the one hand, such campaigns would be ineffective because Western organizations did not have access to the large rural populations that contributed most to high fertility. On the other hand, such efforts were liable to stir up anticolonial and racial "sensitivities." Nor could foreign groups "expect to work through native 'fronts.'" Instead they needed to cultivate "indigenous understanding of the problem and dedication to its solution." "The most important generalization to be derived" from the discussion of "methods and means of influence," Dudley Kirk noted, was "that there are no tricks of manipulation that are likely to work in this field. Whether or not we wish to do so, it would be extremely difficult, if not impossible, to manipulate and control mass opinion from the outside and to bring it to bear on leadership."[169] Instead, "the best approach, and perhaps the only potentially successful one, is to play it straight."[170] To do so, the committee drew on and contributed to innovation diffusion theory, which originated in the communications field. This theory suggested that the place to start would be to persuade the educated and informed publics.[171] This communication strategy follows directly from the sociological conceptualization of cultural change that undergirds transition theory.[172] Diffusion of scientific facts and family planning knowledge to persons of influence would be most effective in persuading political leaders to persuade the masses. Specific steps to cultivate local leadership included "plant[ing]" articles on population matters in the local press directed to elite audiences. Additional steps included publishing "cheap editions" of scientific materials on population and, of course, encouraging "Council fellows" and "indigenous professional colleagues . . . to write articles for professional social science journals; to make speeches to professional and intellectual groups; and to promote discussion in teachers' meetings and comparable forums . . . and on radio and TV."[173]

Once concerns were roused in the literate public, the council could provide the information and materials to those indigenous individuals to use to advocate for lower fertility rates, in keeping with their own cultural standards and "custom morality." This way the council could effectively stoke genuine commitment by local leaders to engage in a mass campaign without inflaming local racial, anticolonial, and sexual sensitivities. This approach was also said to be consistent with Notestein's earlier conclusions that in the West and Japan the adoption of family limitation came about through "private word-of-mouth, rather than propaganda."[174] Moderate, reasonable, and scientific, this approach would "present the problem, not propagandize a solution."[175] In the committee meetings, council leaders often spoke as though there were no existing indigenous birth control or family planning organizations of any importance in the Global South. Thus they positioned themselves as the subjects possessed of superior knowledge and the methods to disseminate it to as yet unenlightened communities.

Following from this model of public relations, the council's principal initiative was a fellowship program that brought indigenous scholars from former colonial nations to complete training in demography in US universities and research centers. In the course of its first twenty-five years, the program supported 994 fellows, 710 of whom were scholars from the Global South. The fellows studied in demography or biomedical programs at universities such as Princeton and the Universities of Wisconsin, Chicago, and California at Berkeley.[176] Dudley Kirk later claimed that "over a third" of "the leading demographers in less developed countries had been Population Council Fellows."[177] The council also founded regional centers in conjunction with the United Nations that supported demographic knowledge building in Asia and South America.[178] Thus, the fellowship program and research centers produced local, indigenous expertise trained in research "invented here" who could influence policy in many elsewheres.[179] And as Irene Taeuber noted in 1954, "The responsible people in these areas who have western training know the population problem in considerable detail—and they know the problem not as a theory but as a nightmare."[180]

How much the committee served to provide scientific legitimacy to preexisting positions held by Population Council leadership and staff and how much of the council's approach was shaped by the expert opinions it solicited is difficult to discern from the records. The invited experts on demographic and population matters were by no means independent agents. The population scientists were all closely tied to Osborn, the council, and the OPR. The communications scientists and religious experts were also drawn from the social worlds of Princeton and the Rockefeller Foundation.[181] However, the committee's process—calling meetings, soliciting expert opinions and recommendations, entertaining opposing views, and publishing findings—performed the careful, moderate, and scientific character the Population Council claimed for itself. It provided a means with which to persuade friends, competitors, opponents, and especially funders that the council deserved authority for matters of population policy and family limitation programs. Thus council leaders both encouraged the view that only moderate groups should be funded and successfully demonstrated that they were such a group.

Theirs was a successful strategy. Between 1958 and 1968, the council garnered $41 million in support of demographic research worldwide.[182] Those funds facilitated the global reach of the council's program of technical assistance and quiet persuasion, enabling it to have enormous influence on population policy, family limitation programs, and the field of demography. Beyond the fellows program, the council supported the aforementioned demographic research centers in India and Chile, the IUSSP, and population conferences jointly sponsored by the IUSSP and the United Nations in 1954, 1964, and 1974.[183] Under the auspices of technical support, the council set up "advisory personnel" in Pakistan, South Korea, Thailand, Taiwan, Turkey, Sri Lanka, Iran, Tunisia, Morocco, and Kenya. The council provided grants to India, Pakistan, Korea, and Taiwan for family planning demonstration trials and population planning.[184] It also funded studies on contraceptive knowledge, attitudes, and practices, known as KAP studies, in these and "more than a score" of other developing nations in conjunction with family planning demonstration clinics.[185] Through these studies, demographers were able to demonstrate that "some motivation for restriction [was] present," and thereby they were able to persuade governments to initiate family planning programs.[186]

The council's approach relied on the rhetoric and practices of public health, particularly maternal and child welfare, to represent family limitation as a means to secure a better living. But as with demographic transition theory, the committee's rhetoric rendered the status of women and children a marker of the achievement of bourgeois cultural competencies.[187] The well-being of individual women was secondary to the goal of containing the growth of "other" nations. These commitments became manifest in the 1960s and 1970s when the council championed precise statistical targets for fertility reduction and the IUD for large-scale family planning programs designed to meet those targets.[188] The council played an essential role in the development and distribution of this contraceptive through grant-supported research beginning in 1961.[189] By 1968, Frank Notestein claimed, "seven million women [had] been fitted with IUDs and they were being manufactured in South Korea, Taiwan, Hong Kong, India, Egypt, Pakistan and Turkey" at a "cost of less than three cents each."[190] The Population Council endorsed the IUD because it was said to be highly effective in the face of weak motives.[191] In contrast to the hormonal pill being developed by Planned Parenthood at the same time, the IUD did not require sustained motivation to prevent pregnancy.[192] It only required initial "acceptance."[193] It was "birth control for the nation."[194] At the same time, through its domination of the demographic statistical apparatus, its demographic figurations of the population problem as "a Frankenstein," "a nightmare," and a potentially preventable catastrophe, the council stoked the mathematical panic that moved nations to initiate intensive population control programs.

• • •

By the late 1950s, the population explosion and the demographic transition had "achieved the status of a certainty in the social imaginary."[195] Hugh Moore's pamphlet, *The Population Bomb,* circulated widely. Population Council publications told a similar story of "too rapid" growth in the monotone of social science. Demographers such as Kingsley Davis published editorials and features warning of the population explosion in national venues such as the *New York Times*. Transition theory circulated widely in both scholarly circles and the popular press. College students read about the demographic transition in Davis's textbook, *Human Society*. In 1959 CBS broadcast its one-hour *CBS Reports: The Population Explosion,* hosted by Howard K. Smith; the show garnered 9 million viewers. The CBS report was repeated in 1960 with an extra half-hour on how population impacts economic development. Both reports focused on the situation of India and the need for fertility reduction through family planning to rebalance the demographic scales.[196] As transition theory and the figures that proved its predictions circulated, the sense of emergency and panic mounted. Before the war, the explosiveness resided in the political consequences of population pressure; now, population growth itself was explosive.[197] With descriptions of Frankenstein populations, explosive growth, and nearly limitless capacities of fertility, how could informed and educated publics not feel the urgent need to act?[198] The numbers proved that the danger was all too real. Thus, although demographers decried the emotionalism of birth control advocates, demographic figures conveyed their own intense affects, amplifying fears about a monstrous future.

The demographic transition was (and is) the standard description of how and why population changed in the modern era. However, it does not accurately describe the historical process of demographic change in Western nations, on which its narrative of change is based. In France, fertility rates declined before any substantial industrialization occurred. In England, fertility rates fell first and most significantly in rural areas. Likewise, the secondary premise—that rapid population growth would undermine economic development in Third World nations—was not substantiated. Even now, as demographers acknowledge, "social science research on the precise relationship between population growth and economic development remains inconclusive."[199] Moreover, even as the model was being developed, the United States and Europe were in the midst of a rise in fertility that produced rates rivaling those deemed abnormal in the Global South.[200]

But the theory's persuasive power did not rely primarily on the referential status of demographic data. Instead, data gathered and analyzed in the transition framework provided new mathematical credibility to the gendered coloniality and familiar sticky objects of the Malthusian world. As Dudley Kirk noted in his 1996 review of transition theory, "It provide[d] a general historical model of how all current and emerging colonial and non-European societies could be placed in rank order, and an evolutionary typology constructed."[201] Graphic representations of the demographic transition show the progress from primitive to modern. Europe's mythic past continued as the normative case against which the deviance of developing nations could be elaborated: they were where Europe was generations ago, and Europe's example provided lessons for them to master. Moreover, transition theory represented the "Third World" as out of step and out of balance because their declining mortality was premature and their declining fertility was belated. Demographers explicitly embraced a liberal perspective on race, but biology was still an invisible cause of the differential fate of nations. It was just displaced from race to gendered differentialism. The conceptualization of natural fertility makes women's bodily processes the source of the procreative excess assumed by the demographic transition. The biological "fact" of women's excessive reproductive capacity combines with different cultural heritages to organize the modern demographic figuration of population change. How well societies managed and contained the natural excess of women's bodies indexed their stage of social evolution, demarcated in the phases of the demographic transition. The theory thus encompassed a diversity of people and places within a universal scale that recognized and disavowed difference on a gendered axis. Regions represented as the areas of greatest potential growth were routinely represented as feminine. In this way, the hegemonic masculine logic of the theory gave the alibi of nature to its narrative of the fate of nations. It distanced demography from the racial bias associated with eugenics but legitimated the eugenic principle of biomedical interventions to alter abnormal, out-of-balance, and out-of-control fertility.

At the same time, transition theory underscored the necessity of normative sex roles within the small-family ideal, the gendered restraints of functionalist sociological theory in which the family, composed of the individual, his wife, and his children, were the cornerstone of democratic society. The small-family ideal required fertility management as an essential cultural competency of modern life. Developing nations that adopted the small-family ideal could become nations of modern neo-Malthusian men: individualistic, empirical, achievement oriented, acquisitive, and future focused, with educated wives and striving children. Demographic transition theory elided the gender politics that shape reproductive processes, contraceptive practice, and population dynamics, even while tracking women's bodies. There is no discursive space within it in which to observe or theorize women's individual sexual or reproductive interests, motives, or constraints. It simply reiterates the modern masculinist heteronormative sexual narrative. At the same time, it represents women of "other" cultures as subject to traditional patriarchy, universal early marriage, and endless childbearing as their sole source of prestige. The status of women continued to be read as indicative of social development, reaffirming the West's conceit that it treats "its women" the best. It simultaneously recuperated the other, "downtrodden," women who were a staple figure of colonial discourses. The most dangerous of these was the "as-if" woman of natural fertility, generally represented as the rural village woman.[202] Her longer reproductive career threatened a greater excess of "functionless reproduction." But demographers were not concerned with rescuing her as much as managing the danger that her uncontrolled fertility represented to the world. Women of the Global South were once again positioned as objects to be feared and contained.

Finally, demographic transition theory provided the grounds for building the field. Resources flowed to the scholars and institutions that developed the theory, enabling US demographers to define the dominant research agenda of the field throughout the mid-twentieth century and to dominate the recruitment and training of demographers internationally. Through their work, the OPR and the Population Council peopled government statistical agencies, universities, research centers, and private agencies with experts trained to conduct and interpret research through the neo-Malthusian framework "invented here." In consequence, demographic discourse effectively shifted attention away from the question of economic relationships between the developed and the developing world. The failure of Western nations to fulfill their pledge to invest in development projects is invisible in the landscape of the demographic transition.[203] Instead, it reinscribes excessive propagation as the primary cause of poverty. The issue was not a simple matter of a "fair deal" for the "Third World," but of modernizing family form and function. Less developed nations had to do their part by constraining their fertility and thus their growth so that whatever economic development aid the West provided would pay off. Such fertility management programs became a requirement under the Johnson-era foreign aid programs.[204]

That population growth would handicap economic development was taken for granted as common sense and scientific fact.[205] The statistical grounds of that fact, however, were not produced until the end of the 1950s, more than a decade after the transition theory and the population crisis predicated on it were disseminated worldwide. The statistical proof came in Population Council–sponsored research conducted by Ansley Coale and Edgar Hoover on the population and economy of India in the 1950s. The figures they produced and the process of their production is the subject of the next chapter.

6

"Second Sight" and "Fictitious Accuracy to the Numbers"

> Population, therefore, is not a question of the relation of numbers to food or income per head, but of the adjustment of numbers to a social purpose. It is a problem of social ethics, a problem of the standard to which the community wants its numbers to live for and live up to.
>
> —GYAN CHAND

ALTHOUGH THE APPEARANCE OF RAPID POPULATION GROWTH PRECIPITATED a sense of crisis in and about many regions of the world, at its founding meeting, the Population Council identified India as the location where the population problem was particularly severe. The council subsequently invested considerable effort to encourage population control there.[1] The council's plan targeted its public relations efforts to India's "informed and educated" public, provided demographic training for indigenous scholars, and gave technical assistance for family planning trials and demographic analysis. It assumed that the "true" nature of the population problem was unknown in emerging nations and that dissemination of the population facts was necessary to generate public support for state-run population control programs. But the expectation of ignorance fundamentally misconstrued the state of the Indian population debate and birth control advocacy. In the late colonial period between the two world wars, the "informed and educated" Indian public became well versed in the population principle and the subcontinent's purported population problems. A strengthening nationalism and British concessions to local rule fostered the growth of a vigorous (largely urban, middle-class, and upper-caste) civil and political society, in which discussion of India's future, its development, and modernization predominated.[2] Those debates took on greater urgency as "anticipations" of "decolonization and independence" increased in the 1930s and 1940s.[3] The 1931 census in particular generated voluminous commentary among colonial administrators, Indian social scientists, and nationalists.[4] With the further devolution of governance to the provinces in 1935, planning for postindependence development constituted a central component of nationalist politics.[5] The question of India's population, its size and trends, figured prominently in those debates, both in the colonial discourse in which population knowledge was configured and in the nationalist planning that translated that knowledge in the interests of the Indian nation.[6]

Birth control also circulated as a topic in these years within both the elite and popular culture.[7] Commercial contraceptive products, both domestic and imported, became available in the early 1920s.[8] The first Indian birth control clinic opened in 1927 in Bangalore.[9] The All India Women's Conference voted to endorse birth control instruction in public clinics in 1932. The All India Medical Conference passed a resolution in support of contraception in 1935, two years before the American Medical Association passed a similar resolution.[10] In addition, birth control discourse circulated globally between metropolitan and colonial worlds in a multidirectional exchange. For instance, *Marriage Hygiene,* an international sexology journal recognized by Western sexologists as a leading journal in the field, was published by A. P. Pillay out of Bombay. The journal, which included articles by authors both in and outside of India, regularly carried articles on contraceptive methods.[11] At the same time, accounts of birth control advocacy and clinical efforts in India circulated in British and American birth control publications.[12]

A topic of scholarly discussion throughout the early twentieth century, the discipline of demography also coalesced in India during the 1930s as Indian population scientists translated the official census data, conducted independent research, and participated in the international population discourse. In particular, Indian economist Radhakamal Mukerjee played a prominent role in international population circles, championing migration as a solution to global misalignments of population and resources and challenging the global color line produced by immigration restriction in Western nations.[13] Of greatest importance, Indian intellectuals and academics participated in formal planning processes of the Congress Party, which drew broadly on the expertise of the nation's scientists, both before and after independence. As a result, population discourse was woven into founding documents of the new nation.[14] With independence, India became the first nation to articulate a population policy and to incorporate family planning into its portfolio of national health programs. Thus, the "informed and educated" Indian public and government targeted by the council were already quite current on population matters.[15] Nonetheless, the representation of India's "backwardness" dominated mid-twentieth-century demographic discourse in the United States. That discourse repositioned India as a dependent client of Western science as it built population figures that identified India as "always already overpopulated" and its family planning program as "an always already failed project."[16]

This chapter investigates the configuration and circulation of demographic representations of India's population in the mid-twentieth century, juxtaposing the Indian nationalist configuration of the population problem with that of US demographers. It begins by sorting briefly through colonial representations of India's population growth to sketch the context of anticolonial counternarratives produced by Indian social scientists, most of whom were Western-educated upper-caste men with strong nationalist commitments who were engaged in both local and global population discourse. The boundaries between population science and birth control advocacy were not as sharply drawn in India as in the United States, and thus their birth control advocacy is discussed as well.[17]

This is a complicated story. It involves interventions into a discourse fundamentally shaped by the colonial state and conducted through/on its terms. The areas of concurrence and divergence between Indian nationalist population science and the Western gentlemen's club of demography epitomized Indian scholars' predicament as unwilling colonial subjects. Thus, scientific knowledge, deployed in both colonial discourse and Indian counternarratives, served as a mark of modernity for all concerned. In particular, the health of the nation, judged by mortality and fertility rates, was an important site of contestation over who could best fulfill the governmental obligation to "look after the well-being of its subject peoples."[18] As a result, as with so many other issues of national concern, the population question and family planning program were shaped by ambivalences about modernity and its meaning in/to/for India. Indian academic discourse and nationalist planning were framed by questions of whether or not India was already overpopulated, would become overpopulated, and what the consequences of population growth might be for national well-being. The analytic lens employed here follows Benjamin Zachariah's observation that "if such intellectuals ended up by legitimating" dominant discourses, "it is not always to their intentions that we must look . . . but perhaps to the logic of the situations they found themselves in."[19] Thus the chapter emphasizes their use of modern science for renegotiating colonial representations of "abject" India and illuminates moments when their commitment to care for the people is betrayed by the underlying class-caste paternalism in their accounts of the people.[20] It illuminates aspects of masculinity politics that brought Indian population scientists into concurrence with Western population discourse, even as their perspectives on the population problem were marginalized within that discourse.

The chapter also examines the circulation of American demographic discourse about India and its consolidation into a discourse of detrimental growth in the 1950s and 1960s. It provides a close reading of the numerical inscriptions in two enormously influential texts that helped solidify the representation of India as the exemplar of the population explosion. The first book, Kingsley Davis's *The Population of India and Pakistan,* raised the alarm about the "explosive" trends in Indian population dynamics.[21] Davis's figures were cited by Notestein in his early renderings of the demographic transition as well as by the influential Bhore Committee Report on Health in 1946; they were subsequently incorporated in the independent government's first five-year plan.[22] The second book, Ansley Coale and Edgar Hoover's 1958 *Population Growth and Economic Development in Low-Income Countries,* constructed projections of the likely course of population and economic growth in India that were taken up in development discourse as the factual proof of the deleterious impact of rapid population growth on economic development.[23]

Although in both cases their research drew upon sources of information and research conducted in and by Indian demographers, statisticians, and birth control supporters, the narratives of population change in American demographic figures represented India as a dependent client of Western science, a nation where the rational control of fertility had yet to be taken up "to any appreciable" degree.[24] As the plans of the Population Council's Ad Hoc Committee on Policy had done, the Davis and Coale and Hoover texts obscured the ongoing engagement of Indians themselves with the population question. This is the case even though they acknowledged that India was the first nation in the world to establish a family planning program and a population policy.[25] The chapter repositions Western representations and Indian discourse and policy within the prism of a gendered geopolitics of knowledge. Thus, by resituating midcentury US demographic narratives within the landscape of Indian scholarly and governmental activity before and after independence, the chapter is also an intervention into the conventional view of India as behind and belated with regard to population and its management.[26] In fact, it shows the manifold ways in which India was in the vanguard of population control. The chapter closes by assessing the weight of midcentury population figures on the trajectory of Indian and global population politics in the 1960s and 1970s.

The Population Discourse in India: "The Testimony of Figures"

As the nationalist movement came to fruition at the end of World War II, leaders of the new India sought to throw off the "colonial yoke" and establish the nation's autonomy and self-reliance.[27] Mostly Western-educated upper-caste men, they would modernize the nation by coupling the power and resilience of India's rich culture and traditions to modern science. A principal goal was to raise the standard of living of the nation's citizens. While framed in Malthusian terms of the balance of population and resources, the goal of improving living conditions was generally stated not in dry figures of per capita income but in lofty sentiments of supporting a life worth living. As Gyan Chand, a prominent economist who would advise Nehru's government, stated it, the goal was to establish "a level at which life can really become creative, at which it ceases to be an unremitting struggle for existence and provides the opportunity and scope to participate in a broad-based culture."[28] India's leaders looked to the successful example of the Soviet Union, whose progress since its 1917 founding seemed miraculous.[29] India's modernization would also rely on a series of five-year plans, the first of which commenced in 1951. That plan included the world's first national population policy, the formulation of which reflected the tropes and tensions that had framed population discourse in British India.

The problem of India's population featured regularly as a trope in colonial discourse. This is not surprising when one recalls that Malthus worked for many years training colonial functionaries at the East India Company's college.[30] As Malthus prescribed, population statistics were employed to tell the story of the masses, whose history was otherwise unrecorded. For the British, this was a self-serving story of Indian inadequacy. Population figures provided evidence of the subcontinent's lassitude, which resulted from/in the triumvirate of "filth," "poverty," and "backwardness." The result was an "unruly" and "ungovernable" populace.[31] As David Arnold notes, "Census data and reports were often used to chide Indians for their subordination to nature and imperfect hold on modernity." Thus, one prominent public health official argued that Indians needed to be educated in the "hard facts of existence," shown "how the people of other countries have solved their problems," and then be left "to choose . . . how they will plan their lives."[32] Following Malthus, census reports described Indian population trends as the work of nature, not reason. Recurrent famine and disease were said to be unimpeded by Indian society and thus kept the already excessively large population in check.[33] Census commentary also recurrently expressed concern that British rule might upset the natural balance of India's population. For instance, the Bengal report in 1878 noted that "British rule has established peace and security throughout the country, and so far has removed some of the causes which were at work to check the natural increase of the people."[34]

Thus, British officials represented themselves as benevolent rulers. They claimed to have brought stability to the subcontinent while leaving Indian society's most cherished customs and traditions undisturbed.[35] The narrative of Indian unruliness and the regime's "polite restraint" provided an apology for the Raj's inability to secure the public welfare, a basic responsibility of government. Instead, it blamed Indian climate and culture for the sad state of the health and well-being of India's population, eliding the devastation caused by British colonial practices.[36] The specter of overpopulation, likewise, helped resolve the "paradox" of the regime's claimed beneficence with the reality of the increasing impoverishment of the Indian people: the fault lay in the natural excess and lack of prudential restraint on the part of India's masses. Finally, it also justified the regime's crisis orientation, in which it acted to protect its own first and always, intervening in the wider society only when things got so bad that "humanitarian feeling" overrode restraint.[37]

As famine and epidemics receded and population growth began to be registered in the early twentieth century, the imperial government expressed intensifying anxiety about India's population growth. The question of whether India was overpopulated was increasingly answered in the affirmative. Census reports in 1921 and 1931 framed the matter as a "race between population and India's ability to produce food," a race, they concluded, that Indians were in grave danger of losing. On the one hand, they concluded that population growth suggested the benefits of colonial administration in reducing the effects of natural checks, especially famine.[38] On the other hand, it concluded that growth put greater strain on the already limited resources produced by India's so-called archaic agriculture, and thus limited its ability to prosper from the benefits of colonial rule. The 1931 census report declared, "The increase of population is from most points of view a cause of alarm rather than for satisfaction."[39] The colonial administration, while suggesting that contraception was urgently needed, took an official hands-off policy, again taking a stance of polite restraint in social matters. Yet, although officially neutral, colonial authorities permitted contraceptive imports and advertising, and public health and census reports praised Indian birth control advocacy.[40]

In this context, where Malthusian logic defined the terms of securing the public good, how did Indian intellectuals understand the subcontinent's population dynamics? How did they respond to the "brusque reminders of their pathetic frailty in the face of nature's Malthusian assault" with which census figures and commentary "bombarded" Indian publics? Indian nationalists countered that British policies were responsible for the impoverishment of the populace because they caused great harm to the nation's economy and the well-being of its people. In their view, the failures of famine relief and epidemic controls proved that colonial officials did not and could not adequately care for the people, which bolstered their demands for self-government. Indian economists and sociologists, too, drew broadly on census and public health statistics[41] as they "translated" census figures into an argument for a strong independent India.[42]

As Gyan Prakash has argued, Indian intellectuals in the late colonial period deployed science in their contestation with British officialdom. Science became a "second sight," "a space opened for the subjectivity and agency of the Western-educated indigenous elite."[43] With this vision, they represented India as degenerated from a state of past glory, a glory in which ancient Hindu science anticipated the principles of modern science. Composed in the long shadow of colonial discourse, the rebuttal by Indian intellectuals attempted to renegotiate the terms and articulate a vision of India's future that combined the power of science and the ancient heritage of Hinduism.[44] Under the leadership of such intellectuals, with their knowledge of both sacred texts and modern science, India would be regenerated, restored to greatness in a particularly Indian modernity. Indian population science followed the pattern Prakash describes. For example, Mukerjee claimed in the first sentence of his 1938 book on food planning "that the ancients comprehended" the population principle, citing the *Brihad Aranyaka Upanishad*.[45] As his biographer notes, Mukerjee argued that "the understanding of India will suffer from fundamental weaknesses and distortions if an attempt is made to interpret the Indian social situation in terms of Eurocentric concepts derived from the pre-industrial stage of European socio-economic evolution."[46] The use of English economic models had, in his view, resulted in "rural unsettlement and a decline in agriculture."[47] His research developed solutions specific to the Indian context and Hindu principles. In particular, he rejected individualism in favor of what he saw as the tradition of "'communalistic' institutions."[48]

Indian commentary on population matters unfolded in dialogue with colonial census figures and was framed by the tensions over who and what was to blame for the deplorable situation of the people. And, David Arnold has argued, the Indian middle classes over time absorbed and internalized "in their own discourses the demonic demography of the colonial census reports." In that sense, it is not surprising that discussions of the population question by Indian economists and sociologists "appropriated Neo-Malthusian and eugenic arguments" of official discourse. But they translated them "to their own vision of a strong Indian nation."[49]

The first Indian publication to address the question of overpopulation, published in light of the 1911 census returns, was P. K. Wattal's 1916 volume, *The Population Problem in India: A Census Study.*[50] Wattal, a trained statistician, worked in the Indian Finance Department and was a member of the Royal Statistical and Economic Societies.[51] The book begins with a faithful recitation of the Malthusian catechism; the opening paragraphs restate the principle of differential growth of population and food, positive and preventive checks to growth. Wattal concurred with the Malthusian logic of "abstinence from improvident marriages," saying that "if the resources of the husband will not permit the expense of rearing children and giving them a decent start in life, common prudence requires that such risk shall not be run."[52] He included extensive analysis of marriage, fertility, mortality, and immigration using the 1911 census reports. Accepting the Malthusian link between population and poverty, he expressed concern about the impact of population growth on efforts to raise the standard of living and to ensure an adequate food supply for the population.[53] However, Wattal rejected Malthusian restraint as a preventive check, noting that it "puts too great a strain on the individual." He endorsed raising the age of marriage, but asserted this was not enough. In very careful language he countered objections to artificial limitation, noting that "civilization in every phase is a process of subduing Nature and not of obeying it."[54] In this way, he argued that population limitation should be a matter of public policy. Fellow Indian intellectuals praised his analysis and conclusions.[55]

Although they accepted Malthusian principles as applicable to India, in general, Indian intellectuals rejected representations of overpopulation as *the* cause of India's poverty. As Gyan Chand noted in 1939, "The view that our dire poverty is in large measure due to the growth of our population is not borne out by the facts."[56] Instead, Chand and others identified British imperialism as the "root cause"[57] of "the evils of ignorance, poverty and disease in India."[58] As nationalists, they argued that the cultural and economic distortions of British rule caused India's problems, of which population was one. And they saw the solutions in self-government, championed in the "name of national regeneration."[59] In Mukerjee's view, Malthus was particularly relevant to India, where population pressure on the land was such that small changes in harvests were "accompanied by marked disturbances in the trends of natality and mortality." However, the relationships between poverty, food, and population were "not immutable forces of nature." They were the "heartless course of an ever-widening vicious circle."[60]

Yet, at the same time, these intellectuals expressed deep concern that population growth would hinder efforts to raise the standard of living of the masses, and declared it to be one of the most serious economic challenges the country faced.[61] As Chand noted, "The task of providing for one-fifth of the total population of the world . . . creates for us a task which is as colossal as it is imperative." Yet, he argued, "even if the growth of population is not a cause of poverty in India and attention has to be directed to the adoption of other measures of reform and reconstruction, the increase of numbers is a matter of serious concern if our economic development has fallen short of the needs of our people."[62] He concluded that population growth had to be contained because the level of economic development was so far below what was needed to secure a life of dignity for so many Indians. In an important sense, then, for him, poverty caused India's population problem, not the other way around. Similarly, in his 1938 book, Mukerjee noted that the gap between needs and resources was so great and the imperative of regenerating village life so strong that population had to be contained through adoption of the small-family system.[63]

The discussion of population also participated in the gendered politics of colonialism. One of the population trends noted by the colonial reports was an "adverse sex ratio," in which census reports registered a lower proportion of women than would be expected given natural (that is, Western) ratios.[64] This observation reinforced the hegemonic masculine judgment—Indian society mistreats its women—that animated the colonial civilizing mission. Census reports attributed the high mortality rate among young women to the practice of early marriage and childbearing. Wattal quoted the 1911 imperial census extensively on this point, noting that it showed that high rates of early marriage corresponded to high rates of maternal mortality. The emerging educated classes of Indians interpreted the "deplorable condition of women" as evidence of the degeneration of traditional Hinduism. By this logic, raising the status of women became necessary to regenerating the nation, and middle-class Indians of both genders joined gender-based reform efforts such as raising the age of consent and permitting widow remarriage.[65]

Maternal and infant mortality likewise became a crucial gendered site of struggle over the relative shortcomings of the British government and Indian culture. As Hodges has argued, "Health rather than functioning as the extension of government . . . served as a fault-line that exposed not the reach but the limits of colonial government."[66] Maternal and infant mortality statistics proved the inadequacies of the colonial government to care for the people. The "ghastly loss of child life" served as "an index of the severity of strain to which our people are subjected."[67] Legislative changes in 1919, made in concession to nationalist demands, devolved responsibility for public health to local councils, which readily took up this new responsibility, especially regarding maternal and infant health. But stingy budgets from the regime limited their effectiveness.[68] The nationalist rhetoric on maternal and infant health coincided with the masculinity of the colonial administration but positioned Indian men as the best caretaker of Mother India, expressed as a husbandly protectiveness.[69] "If the wife is not capable of undergoing the ordeal of parturition, the dictates of humanity are loud in favour of moderation. The mother's point of view is very often overlooked in such cases and the result is a shattered constitution and a premature grave."[70] Because the British had failed to secure the health and well-being of women, children, and thus the nation, India needed to establish independence based in masculine nationalism.

Indian intellectuals and political elites likewise took up the language of eugenics in its heyday before World War II. As Chand noted, "Eugenic considerations are as important in India as elsewhere." It was woven into the official census reports to which Indian population theorists responded.[71] But eugenics was often peripheral to the arguments put forward by Indian population scientists. Where India's "empirical eugenics" was present, its biologically inflected language did not simply rehearse Western eugenic thought.[72] For instance, Mukerjee rejected Darwin's individualistic evolutionary logic, arguing that Indian society demonstrated the principle of "co-operation and interdependence" in its evolution.[73] Some argued that caste restrictions on marriage were inherently eugenic and evidenced the scientific underpinning of Indian customs and traditions.[74] But Indian population scientists eschewed the hereditarianism of Western eugenicists. Radhakamal Mukerjee's population theory, for instance, required both sociological and biological components. He argued, “Social biology shows that man modifies and regulates his reproductive proclivities and that prudential restraint and artificial checks operate differently on the birth rates among different social classes.”[75] Chand likewise linked biology and society by noting that “the population of a community is its response to its environment and its size and quality are determined by the nature of that response,” which necessarily involved both the social adaptations to natural forces and “their fecundity and common heredity.” And, he asserted, “social factors are of decisive importance.”[76] Moreover, Chand argued, “no good purpose can be served by introducing the element of race into the discussion of our population problems,” because “India has been a melting pot of races for thousands of years.”[77]

As occurred elsewhere, Indian population discourse intertwined eugenic concerns about population quality with Malthusian concerns of quantity.[78] But questions of quantity were always paramount. Indian population discourse generally deployed eugenic semantics according to Malthusian grammar. Genetic quality was not a central concern. Instead, an idealized figure, the well-fed and well-educated modern Indian citizen, animated calls for limitation of quantity in order to raise quality. That is, the lack of resources, caused by colonialism, not bad genes, accounted for general unfitness in the populace. It had made India “a derelict people” and rendered the masses subject to the vagaries of nature.[79] A limitation of numbers would reduce the strain on resources and thereby would ensure greater health and vitality to the remainder once independence was secured and national reconstruction began.

This is not to say that these elite men did not harbor pejorative judgments about their countrymen. Mukerjee noted that the population problem in India was “due to the illiteracy and ignorance of the economically backward groups.” He also worried about the dysgenic effects of the lower fertility of “literate caste” and “cultured stocks,” which diminished their proportion in the overall population. Following the logic of Karl Pearson, Mukerjee concluded that “the future population will be largely recruited from the culturally inferior classes and communities.”[80] Similarly, to explain differential fertility between Hindus, Muslims, and “the aboriginal tribes,” Wattal relied on Herbert Spencer’s famous dictum that intelligence and fecundity are inversely related. He declared that “the dignity and worth of individual life is at its lowest among the aboriginal tribes which accounts for their prolificness." More hesitantly, he asserted with regard to Muslims that "there is no disparagement of the community as a whole in the statement that intellectually they are not so advanced as Hindus . . . their cerebral development is so much less and as a consequence their fecundity so much greater."[81] Likewise, Chand, citing Adam Smith, noted that "abject conditions are known everywhere to make people reckless in breeding." While he also noted that the assumption of superiority among the upper classes was "based more on social prejudice then scientific knowledge," he concluded nonetheless that it is not "entirely unwarranted."[82]

Such sentiments show how the Malthusian logic of superabundant natural fertility and eugenic differentialist metaphors aligned in descriptions of the "abject" condition of the "teeming masses" without implicating the educated elite's own privileged position.[83] In this sense, population science, like all science, formed a space in which the educated elite "could claim to represent and act upon the subaltern masses from whom they distinguished themselves." The educated elite saw "the subalterns' need for scientific instruction" in all fields, including prudent, scientifically sound reproduction.[84] The Madras Neo-Malthusian League, founded in 1928 by prominent Tamil Brahmin men, exemplified such a stance in its charter, which appealed to "all liberal and advanced thinkers in the country to help the cause and through it, the evolution of a higher humanity in the motherland and the uplift of mankind in general."[85] But as Sarah Hodges has shown, while eugenic rhetoric was common, eugenic actions were not. In the United States, eugenicists secured restrictions such as compulsory sterilization based on social competency and ethnicity; such initiatives were absent in India in the same period.[86] Eugenic arguments supported various marriage reforms directed at reducing "improvident marriage." But the focus was on raising the age of consent and the legalization of widow remarriage, not restricting the right to marry to the able-bodied, as was the case in the United States.[87] Also, solutions to the inadequacies of the "backward" groups that eugenics was called on to support were most often phrased in terms of uplift through education, rather than containment through social segregation and sterilization, as in Western nations.[88] Yet, as Sanjam Ahluwalia has argued, the neo-Malthusian and eugenic inflection of this discourse set Indian family planning on an "oppressive trajectory."[89] In the 1960s and 1970s, when the state acted to limit numbers, the poor, lower caste, and minority communities bore the brunt of those actions. In the late colonial period, however, that outcome was not guaranteed.

Birth control in India, as elsewhere in the 1920s and 1930s, was a matter of maternal and infant health and sexual morality. But in India the boundary between science and birth control advocacy was not as sharply drawn. Indian population scientists generally supported birth control and birth control "propaganda" efforts.[90] R. D. Karve, who studied the topic on a tour of Europe, began to dispense contraceptive information and devices in India in 1921, and experienced job loss and prosecution by colonial authorities as a result.[91] Yet, Indian birth control advocates argued, there were no religious or legal prohibitions against the use of contraception, and they urged public health programs to provide it. Such a program would help India modernize. As Chand noted, conscious control of reproduction only developed in the late nineteenth century. So "we are only half a century behind the most advanced countries in this respect."[92] Radhakamal Mukerjee expressed the matter succinctly: "Man is no longer the breeding animal he once was; he now subjects his multiplication to intelligent care and regulation just as he now controls infirmity and disease by medicine and sanitation." He argued that "rational family planning and education of the masses in birth-control must be accepted as the most effective means of combating population increase." Without it the effect of other measures to improve the standard of living would "be only temporary." He added that the "promoters of the child-welfare movement should definitely adopt as one of their principal objects the broadcasting of practical knowledge about the use of contraceptives and their distribution."[93] Likewise, the Madras Neo-Malthusian League published a journal that advocated dissemination of contraceptive information as a solution to India's population problem.

Advocates urged the adoption of birth control based on Malthusian natural fertility logic, but they also advocated it as a means to reduce the Malthusian "appalling loss of life" recorded in infant and maternal mortality rates. However, they were concerned only with rescuing Indian women from endless childbearing and early death; they did not seek to eliminate the gendered hierarchies of domestic life.[94] Mukerjee argued that the small-family system was part of India's heritage and a common practice before the nation degenerated in the face of subjugation. His vision of family planning involved deliberate reintegration of the small-family system with existing habits and traditions regulating domestic life.[95] Similarly, Gyan Chand noted that "the conscious control" of population depended on the "intelligent actions of individuals . . . [in the] most intimate aspects of their life and must necessarily" challenge "the whole social code which governs sex, family and the place and position of women in society." However, although he characterized middle-class women's "feminist flair as effecting change in marital relationships," that change underwrote a revolution of parenthood in which "the outcome of the individual's desire to master an imperious force for the improvement of his and his children's lot" was paramount.[96] Their perspective inscribed a Malthusian masculinity in which responsible parenthood, limiting fertility to one's means, indexed competent manly modernity. In this sense, they were Malthusian men, sharing key assumptions about the relationship of the sexual and political economy and the value of statistics for understanding these dynamics.[97]

Men dominated the public space and thus the birth control debate in India. However, in separate public spaces, Indian women were early proponents of birth control. In an effort led by Rani Lakshmibai Rajwade, the All India Women's Conference endorsed the provision of birth control in public health clinics every year from 1932 to 1940. These resolutions justified birth control in terms of women's health, infant mortality, and poverty mitigation.[98] The resolutions were framed by the maternalist feminist perspectives of the conference members. For these middle-class educated women, motherhood, indeed scientific motherhood, was central to their endorsement of birth control. It was this perspective that made Sanger a welcome speaker at their 1935 and 1936 annual meetings.[99] Women physicians, who participated in civil society through separate professional organizations often overlooked by men's groups, also championed the cause of maternal health and endorsed contraception as a means to reduce maternal and infant mortality rates. Here again, their goal was to protect Indian motherhood. Women advocates generally did not lay claims to birth control as a means for securing sexual or reproductive autonomy.[100] Instead, as Sarah Hodges has shown, they "strove to construct a normative nationalist sexuality for women," one that "lay claim to a role in the coming Indian nation without disturbing the gendered, classed, and caste status quo."[101]

Birth control was not universally endorsed by Indian leaders. Most notably, Mahatma Gandhi explicitly opposed the use of "artificial" contraception. Although he agreed that population growth had to be controlled, he argued that self-restraint was the only moral method. For Gandhi, national regeneration required self-discipline, and sexuality was a source of corruption—an animalistic force—that undermined it. He promoted a masculinist ascetic ideal of *brahmacharya,* an aspect of which required men to master desire through marital celibacy.[102] In a famous exchange during her 1935–1936 tour of India, Sanger and Gandhi debated the relative merits of prudential restraint and contraception. Gandhi is said to have conceded to Sanger that periodic self-restraint, as in the safe-period method of birth control, was morally acceptable. The most interesting aspect of the exchange, however, is that the two squared off over who knew Indian women's situation best. That is, Gandhi's argument for self-restraint hinged on women taking the initiative to refuse the sexual advances of their husbands. Sanger contested that this was not realistic. Gandhi countered that he wanted women to recognize and exercise their ability to say no. Again Sanger argued that this was not realistic under existing circumstances. Gandhi, claiming status as half a woman, asserted that he, not Sanger, knew Indian women best and could speak for them. Sanger countered that she knew best the dynamics of marriage, and that contraception was required to limit fertility.[103] Indian birth control advocates such as Karve and Chand also vigorously criticized Gandhi's position before and after the meeting with Sanger.[104] Yet, Gandhi's view of prudential restraint would shape the postindependence family planning program, which initially promoted only the safe period method.

Indian population scientists did not actively disassociate themselves from birth control "propaganda." Thus population science and birth control advocacy were closely tied in Indian public and scholarly discourse when the academic discipline of demography coalesced in Indian universities during the mid-1930s. In 1936, the same year the OPR was established, Radhakamal Mukerjee and other prominent Indian social scientists organized the First All India Population Conference in Lucknow. The conference discussed the population research that would be conducted under the auspices of the newly founded Institute for Population Research.[105] In 1937, K. C. K. E. Raja, a physician, produced the first Indian forecast of national population trends, following the methods of

Raymond Pearl and R. R. Kuczynski.[106] The following year, the Second All India Population Conference took place in Bombay. It included a public lecture and exhibition at the town hall, which was a common practice among nationalist groups. Posters about proper nutrition, sanitation, and disease prevention were displayed for the general public alongside charts displaying India's population trends. According to news reports, the exhibit conveyed an urgent message about the need to limit numbers through modern contraceptive techniques and marriage reforms.[107]

The institutionalization of demography coincided with important changes in national governance as a result of the 1935 Government of India Act, which extended greater authority for social policy to locally elected provincial legislatures. With the elections of 1937, the Congress Party resolutely embraced this responsibility, organizing the National Planning Committee, chaired by Jawaharlal Nehru, to lay plans for regenerating the nation and its people. The National Planning Committee drew extensively on Indian scientists to formulate plans for all facets of public life.[108] Although ambivalence toward Western modernity permeated Indian civil society, an underlying commitment to an Indian version of modernity was a fundamental driver of nationalist policy. Britain's failures in securing public welfare had been a powerful indictment of the British and a legitimation of the right to self-rule. Now improvement of public welfare became a national imperative.

The issue of population figured prominently in the committee's agenda. In his speech to the Indian National Congress in 1938, Subhas Chandra Bose identified "our increasing population" as "the first problem to tackle" in the long-term plans of independent India. He sidestepped the questions of whether India was already overpopulated and the methods for containing future growth. He asserted only that "if the population goes up by leaps and bounds, as it has done in the recent past, our plans are likely to fall through." Interrupted by the war and Nehru's arrest in 1940, the committee did not complete its work until 1949. When it published a twenty-six-volume report that year, the population subcommittee report was among them. Radhakamal Mukerjee and Gyan Chand served on the population subcommittee, which explicitly linked population growth and economic development.[109] The 1943 Bengal famine, in which millions died, intensified focus on the population-resource balance.[110] For Chand, containing population growth was essential to "solving the problem of want and misery." Without it "we shall be severely handicapped in our efforts and may fail to make any headway at all."[111] Thus, growth was a liability, making the tasks of national liberation and regeneration all the more difficult. Given the enormity of the task facing the nation, he concluded, "we in India should make stationary population the aim of national policy for a fairly long period." It was essential for "remaking the nation."[112] Mukerjee, who chaired the population subcommittee, concurred.[113]

The first five-year plan of the newly independent India, drawing heavily on the National Planning Committee, began in 1951. The primary goal was to eradicate poverty and it called for the modernization of agriculture and industrialization as the surest route to success. The first postindependence census was conducted in the same year. At the time, it was the largest census ever attempted, and it was a point of national pride that it was conducted according to newly articulated international standards from the UN population division.[114] Hailed as the first enumeration of the independent nation's people, the census was characterized by Indian demographers as an act of citizens' political participation in the new and largest democracy in the world.[115] By this time, Indian population scientists had fully reformulated their lens on the cause and effect of population trends from past causes to future consequences. The question of the relation of population and poverty was now framed as a matter of planning for the future. Radhakamal Mukerjee and Gyan Chand argued that the population's large size and continued growth posed a significant challenge to economic plans. Thus, in 1944, as famine and war continued and an independent future had yet to be realized, Chand noted:

> There is still a fairly widespread feeling that the fear of over-population is an imaginary scare and, if our resources are developed to the utmost, we need have no anxiety regarding our ability to support a growing population. As a reply to the argument that the misery of our people is due to their numbers, the statement has a good deal of sense to it, and there is no doubt that the growth of our population has been unscrupulously used to explain away grave shortcomings of British rule in India. We are concerned here, however, not with the historical causes of India's poverty but its existence as an all-pervasive fact of our national life.[116]

He located the cause of India's population problem not in high fertility, however, but in high mortality. Concurring with Malthus, he concluded that mortality was the "decisive factor in the growth of population in India." It produced a fatalistic attitude toward life, the prospects for the future, and for one's children. The fact that death stalked so many spurred high birth rates because "precariousness of life and the exercise of prudence cannot go together." This recognition undergirded the push for economic reconstruction; the "loss of life . . . constitutes a standing challenge to our national self-respect." Mukerjee concurred and argued that "the chief object" of "birth control in India" was "the reduction of mortality rather than restriction of births."[117]

While there was wide agreement that containing population growth was important for securing national well-being, family planning was only tentatively embraced by the Indian government. The national government remained largely ambivalent about birth control through the 1950s and into the early 1960s. Initially, the government held to a policy of promoting only the rhythm method. In addition, the rollout of funds for clinics and research was very slow, and not all of the money allocated in the first and second five-year plans was spent. In part, this tentativeness reflected caution on the part of government officials, particularly Rajkumari Amrit Kaur, minister of health. Deference to Gandhi's stance accounted for some of that caution, but it also reflected the limited financial means available with which to distribute contraceptives. Provincial governments had some independence in matters of health, however, and some pursued birth control initiatives.[118] The Bombay Municipal Corporation opened free family planning clinics shortly after independence.[119] Moreover, while the funds allocated to family planning by the government in the 1950s were not fully spent, American foundation money poured into the nation, along with a "crowd of Americans" staffing clinics, research projects, and regional training centers.[120]

To sum up, Indian population discourse was braided with aspirations of upper-caste nationalism. It resisted the representation of India's population problem as one of its own making or the cause of its poverty. It focused on the "appalling loss of life" that characterized British India and sought to cultivate a life worth living for the citizenry. These sentiments also resonated with democratic and feminist birth control perspectives. For instance, while expounding the need for "population control," Chand noted that "the birth of babies must remain the most intimate part of the life of individuals and the process continue to be an act of sacrament in the most real sense of the word."[121] For his part, Mukerjee trusted that when "the Indian peasant women" had education, some leisure, and fewer babies who died in their first year, they would readily see the wisdom of limiting fertility.[122]

Yet, there are also clear resonances between the discourse of population change assembled in the international gentlemen's club of demography and the local discourse in India. Like the Population Council, Indian population control advocates noted that a decline in India's high birth rate was unlikely without a concerted effort to spread knowledge of the population problem. As Chand noted, "If that is to happen, a much clearer appreciation of the essentials of the population problem will have to come about and be translated into terms which the common people can understand before it is put into effect in their everyday life."[123] Even as it contested colonial discourse, Indian population discourse characterized the masses as subject to Malthusian natural forces. While they pointed to the British as the cause of India's poverty, they also concurred with the Malthusian economic logic that "whatever measure of success" was made in "mitigating the sordid poverty" of the nation's people, growing numbers would undercut it.[124] But most important, the midcentury population discourse in India was structured by the Malthusian slippage between individual good and the social good, and thereby aligned national population policy to state interests and anxieties. As Mukerjee framed the issue, "Individual happiness and the power of the state [were] equally significant factors in population adjustment."[125] Chand argued, "In bringing up children parents are really agents of society and are entitled to aid and guidance in the discharge of their obligation." The social good of rapid development and the state's authority to guide parents in fulfillment of their social obligation would justify the displacement of individual reproductive agency at the peak of the panic over rising numbers. When the masses are characterized as buffeted by natural forces and unenlightened as to the wisdom of prudent fertility, state intervention is justified as an act of rescue, not a usurpation of individual agency and rights. And it is a deeply gendered rescue, as the masculine state once again acts to protect Mother India from the effects of natural fertility.[126]

While there may appear to be a direct line connecting the underlying undemocratic strains of Indian discourse of population change in the late colonial period to later coercive state practices, it is still important to take account of how the transnational flow of population figures and fears affected the local Indian discourse of population control in the 1960s and 1970s. The readings American demographers gave to the population figures they built and disseminated in the 1950s shifted the register of debates about India's population from academic concern to public panic.

Population Frankenstein and the Melancholy State

Purported to be the first analysis of the Indian population after independence, Kingsley Davis's influential 1951 book set the tone of the transnational discussion of India's population problem throughout the mid-twentieth century.[127] Davis himself forthrightly declared that when he began the study, he "knew little about either India or its population." But he argued that "the stock-in-trade of the foreign observer is always his fresh point of view." His consisted of a "persistently demographic focus" and "comparative and sociological illumination" that would "broaden the frontier of both demographic and sociological theory."[128] While Davis claimed that his was a fresh perspective, the text rehearses longstanding lamentations of colonial discourse both about the data and the people. Davis's analysis positioned the aspirations of the new nation against the standard tropes of colonial discourse and produced a numeric portrait of India's continuing inadequacy.

Davis outlined his task as assessing the possible futures of India given its underlying population dynamics. Davis framed the full significance of the case study, saying, "She [India] has had more usable data than any country of equal backwardness and this fact constitutes one of the strongest reasons for making a special study of her population problems. She exemplifies some of the major difficulties besetting the heavily peopled rural economies of the world."[129] To mid-twentieth-century demographers, India represented an exception to the perennial problem of scanty data about population dynamics in former colonial regions. An uninterrupted series of colonial censuses beginning in 1881 meant that India was a "low income country" in which the demographic situation was "known."[130] From the census records, it was thus possible to build mathematical projections of India's demographic future from its demographic past and present. Identifying India as "a sick region," Davis argued that the data gathered by British authorities "enable us to feel its historical pulse, take its temperature, examine its anatomy, and make a prognosis concerning it."[131]

But, although there were more data about India than about other newly independent nations, Davis still bemoaned the data's gaps. Decennial censuses provided seventy years of snapshots about the population size, age, sex, and caste/class configuration. But these censuses did not record births and deaths, and India's birth and death registrations were even more limited. While episodic reports on epidemics and famines included mortality figures, birth registration was very incomplete and no attempt was made to record marriages.[132] Nonetheless, Davis asserted, "the registration figures can be used for estimates that must come somewhere near the truth, and for the discovery of trends that have some validity and great importance. When combined with census returns, they throw considerable light on basic features of Indian demography. This is especially true of mortality statistics."[133] Davis constructed his analysis by "pooling . . . death registrations with census life-tables," and previous studies by colonial agents and Indian population scientists.[134] Thus the tools and texts Davis relied on to familiarize himself with India were those composed in the late colonial period in debates between colonial officials and nationalists about India's future.

The first sentence of the book announces his commitment to colonial discourse. It states: "The Indian subcontinent has long been known for the gravity of its problems." He went on to describe India's poverty as "so abject that an American has difficulty in comprehending it." This poverty was, in his view, the major problem now that independence and partition had settled its political problems. He further asserted in this opening paragraph, "This poverty would not be a world-disturbing problem if a clear road to improvement could be seen. But the truth is that the way of relief is not altogether clear and certainly not easy, and one of the reasons is the demographic situation."[135] Framed by the logic of gendered colonial difference, the text repetitively reaffirmed the positional superiority of the West as it bemoaned the impossible situation India faced. While he also noted the distortions British imperialism produced, in tone and tenor, the text implies that India's peculiarity is the primary cause of its problems. Thus, he asserted, "It is a sick region, poor and conflictual and ready to fly apart." But held together briefly by the Raj, its data enabled a careful study of its problems and prospects for the future. Davis's narrative of India's population history embraced the temporal distancing characteristic of colonial discourse. Thus, he asserted that India sustained a "thickly settled population" with "Neolithic technology." As the above quotations show, sometimes he personified India as feminine, sometimes his pronouns were gender-neutral. But they were never masculine. His narrative also positioned India as always already overpopulated. Thus, he noted that even fifteenth- and sixteenth-century European "visitors" were "impressed by the density of settlement . . . some . . . considered the country overpopulated at that time."[136]

Still, while excessively large in absolute size, growth of the population was said to be historically very slow. This situation would have continued except for contact with the West. Davis argued that "without this outside stimulus it is doubtful if India's condition would have changed for several centuries." With the advent of "European control," Malthusian checks "lessen[ed]." "Even so, it took a while for European influence to make itself apparent in the population trend." Population thus increased "slowly, though gradually accelerating." Until the first third of the twentieth century, however, decades of population growth in India were "regularly" followed by decades of population loss from famine and/or disease.[137] "At that date [1921] there began an increase which lasted longer and attained more momentum than any previous one, and which is still going on today. In the decade 1921–1931 the increase, almost 11%, was the highest on record for India; and in the following period, bringing us to the present, it mounted to 15%. Thus at a time when the Western nations were beginning to achieve demographic stability, India, with its much larger population, was just starting what appears to be a period of rapid and gigantic expansion."[138]

This expansion potential did not result primarily from a high growth rate. In fact, that figure, 1.2% per year, was comparable to growth rates in the West, which were also calculated at about 1% per year. Instead, "what [wa]s important about India's recently accelerated growth" was "the huge absolute increments to which it [gave] rise." Any sustained growth would produce ever larger increments. To reiterate, the case for explosive growth here was not its rate but the size of each additional increment. Thus he noted the addition of 83 million people in two decades between 1921 and 1941.[139] It was not until the 1961 and 1971 censuses that the averaged intercensual growth rate differed appreciably from that of Western nations. At that point, the growth rate (calculated above 2%) became the index of explosive growth. However, in the 1950s Western population discourse, the tremendousness of India and its population problem were represented in terms of the total numbers added each year and each decade. The addition of those millions was often calibrated by equivalence to small European nations such as Britain and the Netherlands.[140] In this way, India's population history was configured as a case of a (huge) abject other—poor, backward, timeless, and subject to nature's harsh judgment—in contrast to the (small) rationally managed, prosperous, numerate, and progressive West.[141] The positional superiority of Western nations was thus reaffirmed in the portrait of India's excess. Yet India's disorder threatened the West as well—as the explosive growth metaphor conveys.

The major cause of growth that Davis identified was the unanticipated rapid decline in death rates and steady and/or slow decline in fertility rates. As we saw in the last chapter, this pattern of declining death rates was said to result from the importation of modern, Western death control technologies, especially public health measures that improved sanitation and reduced disease. Such actions could control epidemic disease "with virtually no disturbance of private folkways." But for Davis, it was not the effects of colonialism that were primarily the source of India's population problem. The fault resided in "deeply rooted" "mores," code for the traditional gender relations that were said to sustain high fertility. The problem, as he saw it, was that a declining but continued high death rate "strengthened the tendency toward high fertility." On the surface his interpretation of the data coincided with that of Mukerjee and Chand: the combination of declining but still unsettled mortality and the continued high fertility reinforced each other, and led to patterns that were out of balance. But where Mukerjee and Chand regarded the situation as a source of sorrow and damaged national pride, for Davis, it was "a frankenstein."[142]

Drawing on census figures to construct life tables, Davis estimated the age-specific fertility and mortality rates to calculate a net reproduction rate and, with this "intrinsic rate of increase," built projections of the future course of India's population growth. These reconstructed rates, which were analyzed by categories of region, religion, and caste, "enabled" Davis to "predict more accurately the hypothetical consequences of present behavior." With these figures, he drew conclusions that were similar to those of Indian population scientists, but again the tone is very different. Rather coldheartedly, he concluded that the contrast with "advanced countries . . . shows the wastefulness of Asiatic reproductive behavior. A great many children are born, but most of them die. The people of Pakistan and the Indian Union therefore expend a tremendous amount of biological and social energy in obtaining a fairly modest rate of population growth. In this respect their reproduction is like their agriculture, for there, too, they spend much labor in producing mediocre results."[143] Despite this social and biological wastefulness, Davis projected a tremendous increase in absolute numbers. He offered three figures. The first, which assumed the 1921–1941 growth pattern would continue, projected a population of 790 million in 2000. The second assumed that growth would occur at the average speed recorded in the seven census decades, yielding a figure of 560 million in 2000. The third fit Pearl's logistic curve to the eight census totals and projected outward, producing a figure of 635 million in 2000.[144]

He then considered the likely impact of population growth on the nation's efforts to modernize and raise the standard of living. Citing Dutch memoirs of the 1620s, he labeled India a nation of long-enduring poverty and asserted that the standard of living had risen only slightly since then. Eliding the impact of colonial economic relations, he concluded that India had "apparently already reached the point where density and rapid growth are impeding economic development." If the existing population had already undercut development, then future, more rapid growth would further "retard" development.[145] In the first place, he argued, given the large rural population, the addition of more people to the land would likely just add to the already large surplus population. Second, the continued population growth would hinder industrialization in India because population patterns had produced an exceedingly young population, which put a "substantially larger burden of dependency" on the society and economy. "This makes it hard to accumulate the surplus and invest the energy necessary to develop large-scale basic industry," which for Davis was the essence of industrialization. The more rapid the growth, the larger the incremental increase, the more limited would be the resources available to modernize.[146] Following the logic of demographic transition theory, he also pointed repeatedly to the peculiar culture and social organization that sustained high fertility as a hindrance to economic development.[147] That is, the traditional social and cultural patterns purportedly left intact by the Raj had not spurred development in the past, and they were not likely to spur development in the future. He pointed specifically to caste, the joint family, and religion—all features of "traditional" culture—as inhibiting the individual initiative and mobility on which modernization depends.[148] These patterns left "extreme fertility" in place.[149]

These combined factors led him to the "melancholy conclusion" that India's chances of success for improving the standard of living of the masses were slender at best. Still, he concluded that industrialization was the only hope. But, it alone would be insufficient. Also necessary was "a sustained and vigorous birth control campaign." Again he was not sanguine that the government of India would adopt such a full and sustained population control policy. Thus, he concluded, "the demographic situation" would "get worse before it gets better." If the government did not act, then mortality was likely to rise once again because "the demographic account [would] have to be balanced sometime."[150] However, Davis did not provide an exact figure for the impact of rapid population growth on the standard of living and on economic development. That analysis awaited the calculations of another OPR study funded by the Population Council.

Population Growth and Economic Development

Ansley Coale and Edgar M. Hoover's monograph on the economic impact of rapid population growth in "low income countries," conceptualized as a follow-up to Davis's study and published in 1958, was the result of three years of work. Coale and Hoover, who worked at the OPR in this period, specifically credited Frank Notestein with suggesting the "quantitative approach employed in the study" and with "participating in the arrangements that got the project underway." Unlike Kingsley Davis, Coale and Hoover traveled to India in the course of their research. They credited that visit with giving them access to "a large volume of data" and offering "invaluable first-hand knowledge, impressions, and ideas." The Population Council funded that trip and helped shape the research design. Coale and Hoover presented drafts of their findings to the Ad Hoc Committee on Policy in 1956 as they worked on the book, and Population Council staff reviewed and commented on early drafts.[151] In addition, Osborn included their conclusions in his 1958 popular press book, *Population: An International Dilemma*.[152] Thus, the council and the OPR helped produce and position one of the first credible accounts of the quantifiable link between economic development and population growth. The account, framed by Western geopolitical concerns, proved that the population explosion threatened global stability.

The stated aim of the study was to "attempt to give as concrete an answer as possible to a specific question, namely: What difference would it make in economic terms if the birth rate, instead of remaining unchanged, should be cut drastically in this generation?" Their objective was "to bring out the important qualitative differences in economic development resulting from the *choice* of a very rapid population growth or less rapid population growth."[153] To answer the question they built elaborate comparative projections of population and economic growth based on a case study of India. Like Davis, they selected India because it was "an urgent exemplar" of the problem of population growth and economic development in "low income areas." The projections, the core of the analysis, compared various possible scenarios for fertility change and estimated their impact on per capita income changes. Based on those comparative projections, Coale and Hoover concluded, there would be a "large advantage attaching to an *early* reduction in fertility." In fact, they quantified the magnitude of that advantage quite precisely: the per capita income of India "would attain a level about 40 per cent higher by 1986 with reduced fertility than with continued high fertility."[154]

The influence of this work in public policy circles rested in a number of technical and discursive factors. The projections were based on the 1951 Indian census, which was the first census in an emerging nation to use UN standards. The analysis also represented a significant advance in demographic technique. As Frank Notestein's preface states, the "exploration of dynamic numerical models for both the economy and the population" distinguished the analysis "most sharply" from earlier studies. Coale and Hoover singled out "the estimation of the plausible prospective upper and lower limits of population growth" between 1956 and 1986 as "the most quantitative and specific" part of the analysis.[155] Their dynamic population models deployed the key technical innovation of the discipline in that period: component measurement of age-specific fertility and mortality rates, which allowed more precise assessment of the impact of age distribution on population growth. These seemingly mundane statistical techniques configured important numeric elements of the mid-twentieth-century "population crisis."

Their numbers seem to tell a simple and precise story of impending crisis. However, the sense of crisis resulted from the mundane technical decisions Coale and Hoover made in compiling and analyzing the data. This is not to suggest that they engaged in bad science. To the contrary, Ansley Coale was an excellent statistician; the statistical techniques he developed represented the standard in the field and are still in use.[156] The precise numbers, however, were calculated according to the inferential logic inscribed by the demographic transition, which weighed the economic "aspirations" of the new nations against the specter of unbalanced mortality and fertility and found them wanting.[157] Coale and Hoover's figure of a 40% reduction in per capita income provided mathematical "proof" of the Malthusian truism that high fertility causes continuing poverty. The neo-Malthusian anxieties inscribed in the logic of the analysis as well as the sophisticated statistics used gave great critical weight to Coale and Hoover's conclusions, sustaining the representation of high fertility as the principal hindrance to development. Thus the numbers worked as rhetoric to align India's possible futures with the familiar figures of neo-Malthusian gendered colonial difference: excessive fertility, too many people, abject poverty, and failed modernity.

As the first step in building their projections, Coale and Hoover had to establish baseline mortality and fertility rates. However, their primary data sources, censuses, did not provide information about fertility and mortality rates directly. At best, census data provide snapshots a decade apart of total population size and its distribution by age and sex and other categorizations. Therefore, Coale and Hoover had to assemble vital rates through a complex string of estimations that constructed age-specific death rates, the general level of fertility, and a national life table. The fertility and mortality rates that make up their population projections were thus artifacts of statistical processes. Each element in the projections was a complex set of estimated figures. And, most important, fertility rates were the last figure in the sequence of estimations. Thus, they reflect the sum of a "host of inferential technologies."[158]

The calculation sequence began with a calculation of intercensual growth, parsed into average growth per year. This was done by calculating the difference in size between the two census population totals and spreading that number evenly across the ten-year interval. Growth here was assumed to be the sum of the original population plus births minus deaths. Migration is ignored throughout. They next estimated the change in the size of specific age groups between the two censuses. That is, the difference between the size of the population of 0- to 5-year-olds in 1941 and that of 6- to 10-year-olds in 1951 is taken to represent deaths in that age cohort over the decade. Doing this comparison for all age groups enabled them to use standard actuarial tools to build a plausible life table (likely life expectancy for each age group) and estimate age-specific death rates. Once the age-specific death rates had been calculated, it was possible to work backward to estimate birth rates. That is, if migration is ignored, growth is a result of the proportional relationship of deaths to births. Given a specific pattern of death rates, one can estimate the level of general fertility that would be required to produce the resulting intercensual growth. That figure could be used to estimate an annual birth rate. Here again, they used the age-component process to distribute the general level of fertility into the age-specific birth rates. The age-specific rates provided the baseline from which to project future total and age-distributed growth.[159]

Coale and Hoover noted that there were uncertainties in their methods and projections. In the face of uncertainty, they selected what seemed to them to be most plausible. That plausibility was predicated on the demographic transition's characterization of population change in high-fertility/low-income societies. Their projections, therefore, incorporated the differentialist gender logic of the model their case study was intended to prove. Moreover, I contend, their plausible explanations obscured the effects of mortality and amplified the magnitude and effects of women's uncontrolled fertility on growth.

In the first instance, fertility rates in this model were derived from estimated age-specific mortality conditions.[160] So the estimation procedures for mortality had a significant impact on the calculation of fertility rates, and the assumptions made about mortality trends had a major impact on the projections of fertility trends. Take the case of infant mortality.[161] Coale and Hoover concluded that the census data at the youngest ages were least accurate. Age misreporting was a significant issue, as will be discussed in a moment. Here, though, note the impact of their solution to flawed infant age reporting for the reconstruction of infant mortality rates. Concluding that the census-derived infant mortality estimate was too low to be accurate and therefore must reflect undercounting, Coale and Hoover simply picked a higher number that seemed reasonable. They used the figure 225 per 1,000 live births in their projections, although they acknowledged that there was "an insufficient basis for arriving at a very precise figure."[162] Their figure was higher than the official birth registration figure for 1951, which was 125 per 1,000 live births. It was also higher than the Indian government's life table of 185. Because fertility rates in the projections derived from their proportional relations to death rates, the assumption of higher infant mortality pushed up the estimated baseline fertility rates. If the infant mortality rate was assumed to be 200 per 1,000 live births, then the estimated birth rate must have been 40.6 per 1000 population in order to have produced the estimated 1951 population. If, on the other hand, the infant mortality rate was assumed to be as high as 250 per 1,000 live births, then the estimated birth rate must have been 45.7 per 1,000 population in order to have produced a population of the size recorded in the 1951 census (figure 6.1).[163] These higher baseline values were amplified across the twenty-five years of the projections, as a higher estimated initial infant mortality pushed up the estimated baseline fertility rates required to reach the cohort sizes in subsequent censuses.

Age data, especially among younger age groups, were deemed unreliable.[164] But age-specific data were central to demographic projections, so Coale and Hoover developed a separate technique for estimating death rates and the general level of fertility. Rather than using the Indian government's age-corrected census figures to produce their estimated death rates, they relied on the pattern Davis constructed for 1891 through 1941 in combination with stable population theory and comparison to life tables for other high-mortality nations. This means that the correction procedures relied on other hypothetical models of mortality in other similar emerging nations configured by similar assumptions about

VITAL MEASURES ASSOCIATED WITH ALTERNATIVE ASSUMED INFANT MORTALITY RATES, 1951

Infant Mortality Assumed Rate (per 1,000 live births)	*Birth Rate* (per 1,000 pop.)	*Death Rate* (per 1,000 pop.)	*Life Expectancy at Birth* (years)
200	40.6	28.4	34.1
225	43.2	31.0	32.2
250	45.7	33.5	30.2

FIGURE 6.1. Coale and Hoover: alternative assumed infant mortality rates for India, 1951. From Coale and Hoover 1958, 54.

population change, further distancing the numbers from anything that might fairly be termed actual rates. As Timothy Alborn notes regarding actuarial techniques for constructing life tables, "to construct an accurate life table, it is necessary to possess reliable returns for either the reported age or the recorded numbers of births and deaths over successive time intervals—preferably both."[165] Neither was available for India in 1958. They had to be constructed.

There was another factor Coale and Hoover had to contend with. The sex ratios in the census data were far different from expected ratios (112–115 males instead of 105 males to 100 females). As noted above, Indian sex ratios were cited as evidence of the low status of women in colonial discourse, Indian commentary, and nationalist politics, all of which accepted the figures as realistic. However, inferring that the sex distribution in the 1951 census was "inconsistent with a plausible sex ratio at birth," Coale and Hoover simply readjusted it to conform to expected (i.e., Western) sex ratios.[166] Thus they increased the proportion of females and decreased the proportion of males in the 1951 census population. This adjustment brought India's hypothetical fertility patterns into compliance with demographic assumptions about natural fertility. But this decision effectively erased sex-differentiated mortality rates as a factor shaping fertility trends.[167] Coale and Hoover's adjustment of the sex ratio had the effect of increasing the proportion of women assumed to be present in the population at each stage of the projection, which reverberated across the projected fertility and population growth rates. The greater the estimated number of women, the greater the potential growth now and in each subsequent generation. Their analysis thus both amplified the magnitude of potential growth and located the source of growth in women's bodies.

Coale and Hoover's estimations of fertility change in India also built demographic transition logic into the projections, again amplifying the effects of fertility on growth. That is, if growth is taken to be a product of the interaction of births and deaths, then the assumptions about the relative balance of the two are important elements shaping population growth projections. Coale and Hoover drew several inferences about the factors comprising the adjusted-age distribution that again amplified the impact of fertility on growth. That is, an age distribution results from the dynamic interaction of fertility and mortality, as Lotka's formula for the intrinsic rate of natural increase shows.[168] But one cannot tell from the age distribution alone what the relative effects of each element are. Moreover, changes in the age distribution might result from changes in fertility, mortality, or both. For instance, as they noted, the shift in the ratio of children to adults between the 1931 and 1951 census was "consistent" with either "a fairly substantial decline in infant mortality offset by a slight drop in fertility" or "a moderate decline in infant mortality." While they noted that "no published version of the theory of demographic transition states precisely what conditions are essential for a fertility decline," Coale and Hoover concluded that "fertility declines have been non-existent or only moderate" in India.[169]

In support of these conclusions, Coale and Hoover pointed to the demographic facts, such as a stable age distribution between 1931 and 1951. But they also reiterated all the assumptions about social factors affecting fertility decline present in the standard rendition of the transition theory. Their list of social factors included specific assumptions about economies and family formation that were central to postwar social science, including market-based income as a measure of economic status and the subordination of women as a measure of a traditional, custom-bound culture. In their estimation, small changes in the economy or per capita income would not trigger declining fertility in low-income regions. Nor was there a large government family planning effort that would have increased contraceptive availability.[170] They concluded that there was "no evidence of changes in Indian customs or social structure in the decade before 1951 that would have led to a major drop in fertility." Finally, such a decline would not fit with the transition theory's expectations for a nation at India's "stage" of development. On this basis they concluded that India "conforms closely" to expectations that "the demographic transition theory ascribes to a country in the incipient stages of economic development"; therefore, they concluded, it had yet to experience a major fertility decline.[171]

Furthermore, they drew rather pessimistic conclusions about the future, saying it was "questionable whether the economic and social change likely in the next two to three decades" would be "enough to have an effect on fertility."[172] Again using the West as their yardstick, they argued that "the level of economic development in European countries (and in Japan) prevailing at the time that fertility began a significant decline might reasonably be regarded as representing the approximate 'threshold of decline.'"[173] And in their estimation, it was "unlikely that this threshold [would] be crossed in the next two to three decades in for example, Egypt, Pakistan, India, China, Malaya, or Indonesia," all of which were new nations emerging from colonialism. Thus, whatever the specific cause, any appreciable fertility decline in the projection period implied that the Indian transition would have begun at a lower stage of development than that of the archetypical European cases, which they thought was most unlikely. There was no precedent to warrant such a conclusion.[174] "On balance, then," they surmised, "there is little justification for a belief that fertility in India is in the incipient stages of a more or less inevitable decline." These inferences were contrary to the Indian government's own assessment of the census data: that fertility had begun to decline.[175]

However, in their comparative analysis of three models of the effects of fertility change on economic growth, Coale and Hoover's population projections hinge on substantial declines in fertility (figure 6.2). The top line in the graph represents population growth based on 1891–1951 census figures projected out to 1986, with unchanged fertility. The lowest line in the graph represents the projected growth based on a 50% decline in the estimated fertility rate between 1956 and 1986. The middle line represents the projected growth based on a fertility decline that starts in 1966 but still reaches 50% by 1986. They did not pick the 50% fertility decline because it was likely or even achievable; they stated that it was unrealistic. They chose the 50% reduction because a "large contrast" would "make it easier to bring out the nature of the economic effects of alternative population courses." The difference between the unchanged fertility figure and the 50% reduction became the basis for Coale and Hoover's conclusion that India "would attain a level of about 40 per cent higher by 1986 with reduced fertility than with continued high fertility."[176] Based on their estimate, there would be a huge cost of not acting to reduce fertility (figure 6.3).

However, as noted above, they assumed that in the actual course of events, an unchanged fertility rate would be the likely future course. They justified the low fertility estimate with the "assumption that a well-designed and well-executed program of introducing family limitation could reduce fertility by about half in twenty-five to thirty years." They noted that "the basis" for their "alternative extreme assumption" was "somewhat conjectural" and "may be assumed only if the government undertakes an unprecedented, nationwide program designed to introduce family limitation into every Indian village." While they noted the host of problems that would confront such an effort, in terms of resources and scope, they pointed to recently completed attitude surveys sponsored by the World Health Organization and the Population Council suggesting that if such a program were to be mounted there might be "popular acceptance of effective birth control."[177] Nonetheless, the fertility reduction that their analysis was predicated on was, in their own terms, extreme and unrealistic. This is an example of what Gyan Prakash has noted as the unprecedented requirement placed on science in India since independence—to accomplish enormous goals with very few resources.[178]

What is not in this picture is another rate change essential to their mathematical model: a 41% decline in the mortality rate between 1956 and 1976, which profoundly amplified the effects of fertility. Thus, although uncertain fertility changes were represented as the single critical factor in their analysis, the population growth they projected hinged on a certain and continuous steep decline in mortality. By itself, a substantial decline in the mortality rate would produce substantial growth regardless of what fertility patterns were. In fact, in the high-fertility projection the projected growth was entirely due to the assumed decline in mortality. Also, con-

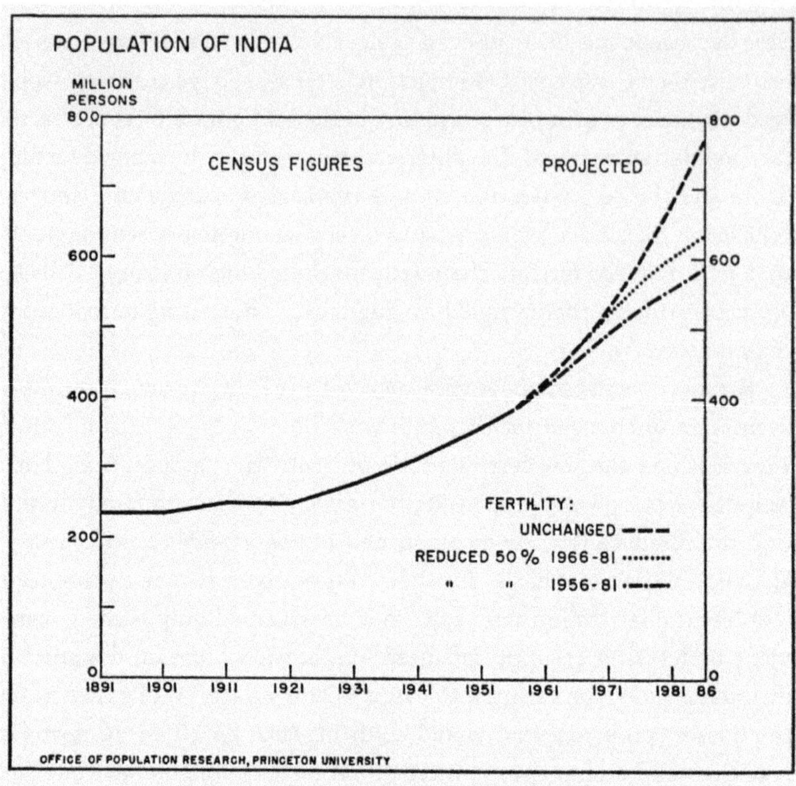

FIGURE 6.2. Coale and Hoover: population projections for India, 1951–1986. From Coale and Hoover 1958, 39.

sistent with transition theory, they built a lag between the mortality and fertility rate decline into the model. Thus mortality rates began to decline in 1951 and fertility at the earliest in 1956 (figure 6.4).[179] This lag meant that even if the fertility rate dropped in the same proportion as mortality, population would continue to grow greatly. This is represented in the low fertility projection where substantial growth was still projected. In the middle case, the lag between declines in mortality and fertility was longer, which led to a longer time period before population growth would stabilize; this indicated that the sooner a fertility decline began, the better.

But although they noted that the projected mortality rates "cannot be accepted as compatible with unchanged fertility" without "at least

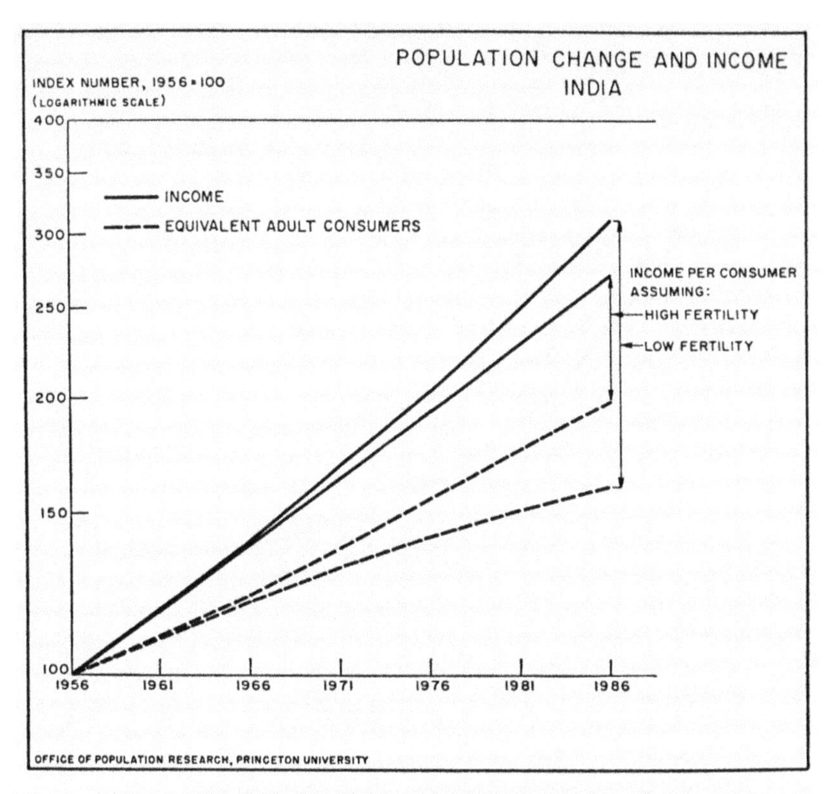

Total National Income, Consumers, and Income per Consumer, with High and Low Fertility, India, 1956-1986.

FIGURE 6.3. Coale and Hoover: population change and income projections for India 1956–1986. From Coale and Hoover 1958, 275.

doubl[ing]" "essential consumer goods," they held the effects of mortality change constant in all three estimates.[180] As a constant, it disappears as an explanatory factor of growth. It is not even in the chart. Only fertility is represented. The effect of mortality on growth was thus obscured and high fertility was read as the singular cause of growth. In this way, fertility and growth quickly became confused and conflated in the population imaginary.[181] This conflation was and is perhaps most obvious in the continued anxious public commentary on higher birth rates among the poor, which, by ignoring the simultaneous presence of higher death rates, creates a false equivalency between fertility and growth rates.

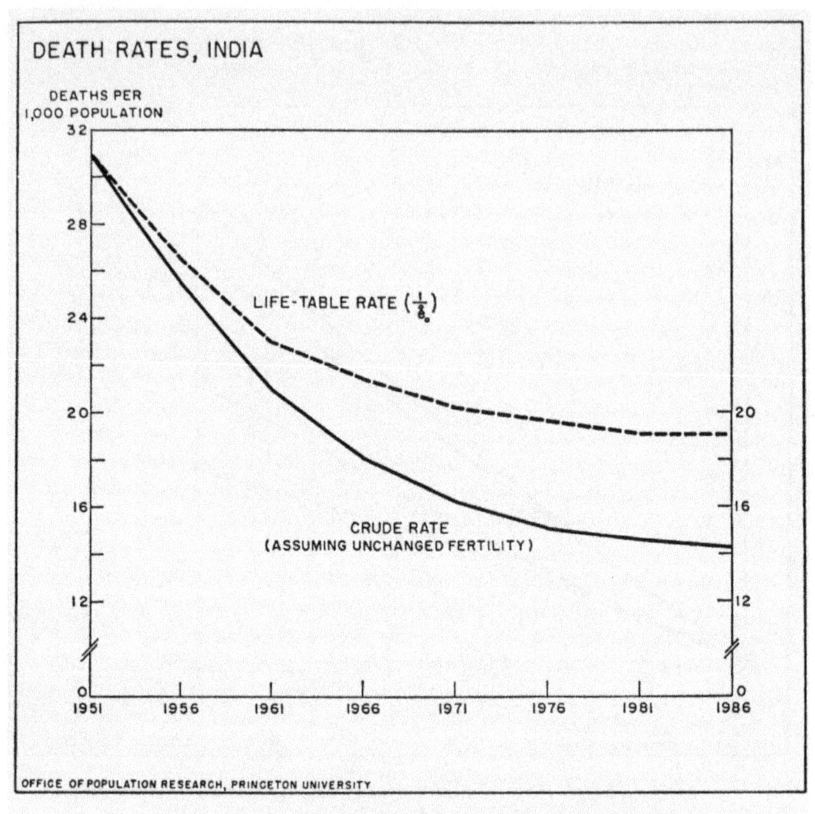

Estimated Crude Death Rate and Life-Table Death Rate, India, 1951-1986.

FIGURE 6.4. Coale and Hoover: estimated crude death rates for India, 1951–1986. From Coale and Hoover 1958, 63.

Demographic logic supports this slippage of fertility for growth, as illustrated by the two differences Coale and Hoover identified "between the populations produced by unchanged fertility and by declining fertility" that were "of greatest importance." In the first, lower fertility would reduce the "fraction of the population at younger ages." This proportion, the dependency ratio, was and is read as indicating the burden on the working-age population of providing for those who are "unproductive." Like Davis, Coale and Hoover argued that expenditures on the nonproductive were "wasteful," diverting needed funds from investment in development.[182] Therefore, a reduction in the dependency ratio would necessarily help a nation achieve greater prosperity, and the only way to reduce the proportion of children in a population and thus the "burden of dependency" was to reduce fertility. The second important difference in the two projections was that the population that resulted from unchanged fertility would after twenty-five years have *"built-in prospects for further rapid growth."* The population resulting from the dramatic decline in fertility would experience a "declining rate of increase" even without a further decrease in fertility.[183] This means that after twenty-five years the growth rates of the two projections diverged further, becoming ever wider, and the fertility of ever-larger groups of women accounted for that divergence. Again, it appears that only fertility makes the difference. But in both cases, the assumption of steep declines in mortality matters for the projected proportion of children and childbearing women in the next generation.

With this model of population change, Coale and Hoover calculated the specific differences in the rate of per capita income growth that would be associated with different levels of fertility. As noted above, they concluded that per capita income would be 40% higher in 1986 if fertility were cut in half. The low fertility projection produced 4.5% annual growth in per capita income by 1986, while the high fertility projection produced only a 3.5 % annual growth rate.[184] There is no denying that that is a dramatic difference. But it is a difference that derives from the assumptions of the model, the hypothetical rate changes, especially in mortality. It is an artifact of the model, its assumptions and estimation procedures.

Moreover, the indicator of economic progress that Coale and Hoover used also had amplifying effects. Per capita income is a measurement of national income that was elaborated at the end of World War II specifically to assess the level of economic development in the Global South.[185] The measurement recast the economies of these largely agricultural and non-monetized economies into the Western terms of dollars a year available to each citizen. Coale and Hoover recognized the limitations of this measure for India, where much economic activity in rural villages was non-monetized. But they used it throughout and explicitly excluded consideration of the value of non-monetized investments in their economic projections.[186] So their measure of value excluded a large proportion of ongoing economic activity and thus undervalued the Indian economy. In addition, per capita income is a measure that is obviously very responsive to population change; it is shaped by the demographic phenomenon it seeks to explain.[187] It thus amplifies the impact of population change on national income. However, its use in Coale and Hoover's analysis was important in another, less obvious, way. Per capita income helped to solidify the shift in the meaning of "overpopulation" from a measure of the density of bodies in a territory to a measure of the density of bodies in markets.[188] To the existing Malthusian anxiety about population density on the land, their analysis added anxiety about the weight of "unproductive" consumption on capital available for investment in development. This is another, perhaps more important, aspect of the neo in neo-Malthusianism.

To sum up, the hypotheticals of Coale and Hoover's model tested not what was likely in the landscape but what was predicted by the transition model. All of the measurements in the analysis were contingent values; they depended on statistical techniques to "fix" bad, missing, and scanty data. Their procedures amplified population growth estimates, fertility rate estimates, and the economic benefits of fertility control, but they discounted the effects of mortality change. Despite their "preaching caution"[189] about the inadequacy of the data on which their analysis rested, the contingencies of their model were lost behind the solidity of the numbers and the vectors of their graphs.[190] The underlying assumptions about mortality, fertility, and economic productivity disappeared beneath the "fictitious accuracy" of their figures.[191] The power of numbers here was, thus, not mathematical precision, but as I have argued throughout, their rhetorical affects/effects in the population imaginary. In this case, their power was evident in the increasing panic over fertility and the credence they gave to the narrative of India's "always already" failed efforts to contain it.

• • •

But were their predictions right? How big was India's population in 1986? These are the inevitable questions whenever population (or economic) forecasts are discussed. However, I contend, whether or not demographic figures accurately predicted the future course of fertility is the wrong question. It diverts our attention. Whether the numbers worked should not be judged by comparing Coale and Hoover's projections of the size of India's population in 1986 to census estimates of India's population size made in 1986. The question of whether the numbers worked is more appropriately addressed to their effects on fertility change in 1958 and thereafter. After all, the point of the population projections was not to get the numbers right about 1986 in 1986. In this sense, they were not intended to be predictive. Coale and Hoover made this point themselves. They acknowledged that their "specificity is not the result of our ability to foresee accurately the actual course of population events in India, but is rather the result of the specific implications of assumptions about population change."[192] Notestein and Davis concurred. As Notestein said, "The hypothetical values . . . are not intended to show what will happen. They were intended to show something of the general nature of the changes to be expected if the conditions assumed actually come about." Further, "if a prediction is accurate, the result reflects luck quite as much as either science or wisdom." This is because "future social economic and political events [would] affect the future components of population change." Notestein also declared that "not much can be said in favor of the realism of such arbitrary manipulations of the future rates of birth, death and migration, except that nothing better is possible."[193] With regard to his projections, Davis noted, "The chance of such projections representing the real future growth, however, is small. . . . The value of such projections is that they show what the region is potentially capable of under fairly normal conditions."[194]

As such statements make clear, the point of population projections was to influence policy and practice in the mid-twentieth-century present. Coale and Hoover's purpose was "to weigh the influence of possible changes in human fertility in India in terms of the *relative* rates of economic progress attainable."[195] The point was to demonstrate the value of a concerted effort by the Indian government to inculcate citizens with the small-family norm and fertility control technologies in order to reduce Indian fertility rates dramatically and quickly. In other words, the point of the analysis was to configure the social relations we call fertility into actionable terms in 1958 and thereby to shape the future. The point, as the Ad Hoc Committee on Policy deliberations made clear, was to provoke an immediate demographic transition in the landscape of the Global South that would approximate the one that existed in the theory.

From that point of view, the numbers worked. Their picture of India's population dynamics was authoritative by 1959. Coale and Hoover's numbers rapidly became a standard reference for those engaged in macro-level politics of development and population control.[196] It "proved" that development efforts would be overwhelmed unless paired with governmental programs to alter fertility practices. And money from foundations and governments in the Global North, along with "a crowd" of demographers and related aid workers, poured in to organize fertility control programs in India.[197] In addition, Indian medical and social scientific experts circulated through US and UN institutes, training programs, and universities to be inculcated with the demographic tools that constituted the neo-Malthusian population imaginary.

In 1961, Prime Minister Nehru, who had long supported family limitation as part of national planning, resisted suggestions that continued population growth made population control more important than efforts to promote economic growth.[198] But the weight of the impact of the flow of Western money and science in shaping the definition of and solution to India's population problem was significant. In the third five-year plan, initiated that year, the government asserted that "the objective of stabilizing growth of population over a reasonable period must be at the very centre of planned development," and the goal of stabilizing population growth was explicitly linked to the goal of doubling per capita income by the 1970s.[199] As Western commitment to foreign aid continued to wane and economic development stalled globally in the 1960s and 1970s, the archive of demographic data registered high fertility as the primary cause of that stall. By that time the ground had shifted at the United Nations. The population problem was acknowledged by all, even birth control opponents, and nations of both the Global North and South supported a resolution to aid nations in dealing with this problem.[200] Within India, rapid population growth may not have been read as the cause of continued sluggish economic growth and a stagnant standard of living, but population control was increasingly seen as an important solution to it. In 1967, with US-trained demographer Sripati Chandrasekhar as the minister of health and family planning, Indira Gandhi's government embarked on a more active plan, setting a goal of reducing the birth rate by 40% by the mid-1970s.[201] Under the leadership of Sripati Chandrasekhar, the Indian family planning program became "incentive- and target-driven," aligning it with the numeric focus of the demographic narrative of the population problem.[202] This was a shift in direction. Gyan Chand had earlier rejected such targeted goals, arguing that for the government to take up such a focus would require the sacrifice of core national values.[203]

In the international arena, India's government continued to represent anticolonial values and resisted suggestions that rapid population growth was the principal cause of poverty in India and the Third World. This resistance is best illustrated by Indian delegate Karan Singh's declaration at the 1974 Bucharest Conference on Population that "development is the best contraceptive."[204] Yet two years later, when population growth estimates continued above 2%, the global oil crisis stalled the global economy, and poor harvests threatened food supplies, the Indian government initiated a "frontal attack on the problem of population" that involved coercive mass sterilization.[205] These seemingly contradictory positions reflect the ambivalences at the heart of Indian population policy. Those ambivalences hinged on the gendered differentialism of demographers' equation of fertility and development.[206]

The current narrative of the 1970s sterilization camps tends to locate the cause in Indira Gandhi's government, her son, Sanjay, and the acquiescence of the IPPF and the Family Planning Association of India.[207] To do so ignores "the conditions under which it was authorized."[208] The longstanding chorus of scientific expertise declared the crisis one "of such overwhelming magnitude, urgency and importance" that it demanded extraordinary action, "a war-footing," and sterilization was the surest method in a region defined demographically by low motivation, high fertility, and high poverty.[209] The numbers stoked the panic directly. The logic of superabundant natural fertility, the elision of the effects of mortality change on growth, and the resulting conflation of fertility and growth haunted this discourse and justified the abrogation of individual autonomy for the greater good of the nation and its population. At the same time, the specter of imperial Malthusianism continued to haunt the discourse as well. Continued high birth rates published in demographic surveys, despite governmental intervention, were read as evidence of the developing world's continuing cultural backwardness, with the reproductive status of women serving as the emblem of its recalcitrance. All the while, the fortunes of demographers continued to grow as their expertise was increasingly in demand.[210] In the mid-1970s, however, the hegemonic masculinity of the demographic narrative of population change would also be challenged, both by postcolonial nationalists and feminists as the population establishment sought to solidify its hold on global population policy through the UN. Those challenges are explored in the next chapter.

Conclusion

Demographic Convictions and Sound Knowledge

> Quantitative and qualitative control of population by a central planning department of the state may some day become a practical possibility. Even if it does, it will probably involve a sacrifice of values which very few people will be prepared to disown.
>
> —GYAN CHAND

I HAVE ARGUED THAT MID-TWENTIETH-CENTURY DEMOGRAPHIC figures should not be assessed based upon the accuracy of their predictions about population trends. The point of such predictions was not to be correct about the future but to spur action in the present. Thus, what matters is the cultural work those figures performed in that present and the haunting effects they have in this historical moment. From that perspective, midcentury demographic figures accomplished a great deal. Produced within the midcentury formation of hegemonic masculinity organized by gender differences and hierarchies, these numbers shored up that formation and aligned family planning to the global designs of those politics and practices. The potent effects of this population crisis problematization were visible in the 1959 report of the US president's committee on foreign aid, generally called the Draper Committee, which included a recommendation that the United States respond positively to requests for family planning aid from developing nations.[1] Thus economic assistance would be contingent on recipient nations reducing their need by reducing their numbers.

In the same year, the population crisis narrative was broadcast to the nation in the CBS special report *Population Explosion,* with Howard K. Smith. Filmed entirely in India, the report laid out the problem as one where the "volatile ingredients are statistical." The report represented the facticity of the statistics with visual reference to the Commerce Department population clock, an odometer-style counter. As viewers watch the dial move, Smith says, "The counter at the top indicates that every eleven seconds the population of the United States increases by one. In the world at large, the increase in population goes on much, much faster than that. In the one hour that this program will be on the air, more than 5,000 people will be added to the present population of the earth." With that image of precisely calibrated growth, the population explosion was described in the standard demographic terms: unbalanced fertility and mortality rates resulted in unprecedented growth rates. Inept and backward others—evidenced by archaic agriculture and the low status of women—were the source of the problem. Smith continues that leaders of populous nations struggled to stave off the dire consequences of unprecedented growth for economic progress and political stability. Moral controversy, rooted in age-old customs, however, blocked effective implementation of the obvious solution—family planning. Of the family planning solution, Smith says that "demographers say a penny spent on family planning can avert more starvation than a dollar spent on more ambitious aid."[2] Thus, the report echoed the Malthusian prudential logic.

Smith interviews a number of Indian officials, scholars, family planning advocates, and average (male) citizens about the problem and possible solutions, including Sripati Chandrasekhar, Lady Rama Rau, Prime Minister Nehru, the prominent economist V. K. R. V. Rao, and Archbishop Joseph Fernandez.[3] While those officials seek to frame the issue as surmountable with sufficient development, and Smith accurately represents India as the first nation to enact a family planning program, the report nonetheless represents India's family planning program as a failure. Presented as a problem of others, in the concluding minutes of the program Smith makes it clear that the report's purpose is to provoke public discussion in the United States about its responsibility to intervene.

By the mid-1960s, family planning programs were initiated with technical assistance from experts associated with the Population Council in India, Japan, South Korea, Taiwan, Turkey, Pakistan, Iran, and Egypt. Demographic studies and research centers covered the globe, and the United Nations supported census development throughout the world. In the same decade, various UN governing bodies passed resolutions that expressed concern about rapid growth, offered support for demographic studies, and "mildly endorsed the idea of [fertility reduction] targets" and family planning.[4] On Human Rights Day in December 1966, a UN resolution championed by John D. Rockefeller III and signed by twelve national leaders declared "unplanned population growth" to be a "great problem threaten[ing] the world."[5] On the eve of the beginning of the UN Human Rights Year in December 1967, the United Nations recognized the addition of eighteen more signatories to the declaration. By 1968, with the publication of Ehrlich's *Population Bomb,* the text with which this book opened, the metric of competent contraceptive practice had intensified to a standard of zero population growth.[6] But as Frank Notestein assured in 1969, "The world's extremely dangerous problems of rapid population growth can be greatly reduced"; demographers had the "prerequisites for a solution in terms of policy, program, methods, public interest, and demonstrated results. Only the work remain[ed] to be done."[7]

But there were limits to the reach and influence of demographic global designs, which became apparent in two international conferences sponsored by the United Nations. A brief overview of the contestation over UN population policies serves to summarize the epistemic and affective alignments of population knowledge. My purpose is not to provide a thorough account of those debates, but to use them to reiterate the gendered geopolitical stakes embedded in demographic figures. In particular, I highlight the moments of alignment between the demographic narrative of population crisis and postcolonial and feminist commitments that continue to haunt transnational feminist reproductive justice politics.

The 1974 World Population Conference

In 1972, the UN Economic and Social Council resolved to develop the World Population Plan of Action, to be ratified at an international conference in 1974 in Bucharest, Romania, which it declared would be the culminating event of World Population Year. Population Council members and associated demographers saw the population conference as an opportunity to "consolidate policy gains."[8] To this end, they participated in preparation for the conference, making substantial contributions to background documents and the Draft Plan of Action. As with most UN conferences, preparatory meetings were held to smooth out differences in advance. Dominated by demographers, family planning program administrators, and related "population experts," these meetings appeared to produce a consensus around the need for an international policy to mitigate the negative impact of rapid population growth. The Draft Plan of Action proposed universal family planning by 1985, and specified global quantitative targets for fertility reduction, including a 10 per 1,000 reduction in crude birth rates by 1985 and replacement-level growth by the year 2000.[9] Consensus on those goals seemed likely based on data from a preconference survey reporting that forty-two countries covering "57 percent of the world population and 81 percent of the population of the developing countries" deemed their population growth rates to be "excessive." Moreover, "34 developing countries" had "policies and programs to lower fertility and an additional 29 provided family planning services for non-demographic reasons." The impact of those programs on fertility rates was not yet clear, although the early success of the Population Council–backed IUD programs in Taiwan and Korea encouraged expectations of substantial declines.[10]

The Bucharest conference, however, did not go as the gentlemen of demography expected. Given the ease with which the conference preparations had proceeded, the US delegation, which included Population Council demographers, arrived at the conference with high hopes that the draft document would be swiftly adopted and the United Nations would give full endorsement to demographic convictions that population control was a necessary principle of international governance. As Bernard Berelson, president of the Population Council, expressed this optimism, "Nothing is so powerful as an idea that has come of age."[11] Instead, the Draft Plan of Action ran into substantial resistance. The principal objections came from the non-aligned movement, called the Group of 77. In the spring of 1974, acting in concert to shape the direction of the Second United Nations Development Decade (1970s), the non-aligned movement secured a General Assembly resolution outlining the New International Economic Order, which called for global economic restructuring to eliminate remaining inequities of the colonial era.[12] On the heels of that achievement, delegates from non-aligned nations, particularly Algeria and Argentina, took charge of key work groups at the Bucharest conference and through them reshaped the Plan of Action to align with their larger political goals. Where the draft declared population growth to be a substantial hindrance to development, and thus a cause of continuing poverty, the Final Plan of Action positioned population growth as a consequence, not a cause, of slow development. As such, the Plan of Action declared, population could not be dealt with in isolation but had to be part of an integrated plan of social and economic development.

With this shift, the urgency of the demographic case for population control ebbed. The Bucharest conference set no targets and did not endorse family planning. It merely suggested that those nations that felt their growth was excessive might consider setting quantitative goals and developing family planning programs.[13] The Final Plan of Action was not the product of the heavy-handed actions of a few birth control opponents, however. Even nations such as India, which had population control policies, national fertility reduction programs, and growth targets, rejected the demographic rationale and goals.

The first week of the conference became so fraught with conflict that US delegates feared the conference would not be able to adopt a plan at all. In the background, however, midlevel officials and technicians worked to craft language acceptable to a majority of participating nations. As the Population Council's report of the conference noted, in the end a plan was adopted because "basic cleavages" were "bridg[ed] with "words that carried sufficiently different meanings to the adherents of the several points of view" to allow the plan's passage by slim margins, paragraph by paragraph.[14] The resulting Plan of Action, demographers concluded, discarded the demographic focus of the draft, pushed population into the background, and ignored the "successes" of family planning.[15] In that sense, the conference slowed the advance of demographically legitimated population control on the international level.[16]

However, the conference did not reject population control itself. It did reject Coale and Hoover's conclusion that fertility decline must precede development. It also rejected the logic that a penny spent on family planning saved a dollar in development aid. But the crisis problematization of rapid population growth was widely shared. The demographic figures compiled in the extensive background documents were not in dispute, only the appropriate response to them. Karan Singh's famous declaration that development was the best contraceptive captures this sentiment. Thus, fundamentally the debate pivoted on the question of who should set the priorities of the UN development apparatus, and what the developed nations owed to nations of the Global South through it. It was thus a debate about the inequitable relations between manly states.

The assertion by non-aligned nations that the population problem would take care of itself with development and the rejection of the demographic case for urgent action essentially relied on the earlier narrative of the demographic transition, in which fertility decline occurred automatically as an effect of modernization. In the aftermath of the conference, demographic commentators said that transition theory had been "transformed into an ideological statement." They represented it and the New International Economic Order as symptoms of what they termed the "Third World syndrome."[17] But, as shown in chapter 6, this version of the demographic narrative follows from the longstanding logic of population debate within anticolonial discourse. That discourse, although worried about growing numbers of the poor and abject, nonetheless resisted the Malthusian logic that population was the cause of poverty, laying blame instead on colonialism and its remnants in international economic relations. In this sense, the framework of the 1974 Plan of Action inscribes a postcolonial narrative of population in declaring that population was a problem that could not be "reduced to the analysis of population trends only" and that population policies "must not be considered substitutes for socio-economic development policies."[18] To the extent that it represented rapid population growth as a problem for member nations, however, it partially aligned with demographic convictions.

Moreover, although the Bucharest conference rejected demographic convictions as international policy, it left considerable room for individual sovereign nations to act to control population growth as they saw fit.[19] The Plan of Action thus enabled nations such as India to resist Western global designs at the United Nations and yet enact aggressive population control measures within their borders. Within a year, in the face of growing political unrest and threats of famine, the government of India would declare a national emergency, suspend democratic governance, and, among other measures, initiate a mass sterilization campaign. Accounts of the 1970s' sterilization camps focus on the elitist motives and actions of the government and the acquiescence of family planning organizations but discount the relentless chorus of demographic voices declaring India's growth to be an overwhelming and urgent crisis that demanded extraordinary action, and touting sterilization as the surest method of fertility reduction in the region.[20] As Adrienne Germain would note later, the mass sterilization policy "was a profound indicator of what can go so wrong if you only look at demographic control goals."[21]

While the main debate at Bucharest centered on the international hierarchy of men and nations, the status of women was also contested. Reflecting demographic convictions, the draft document called for increased educational and employment opportunities for women in order to raise the age of marriage and thereby reduce fertility. The potential demographic impact justified enhancing women's roles, not the potential benefit to women themselves. The NGO Tribune, sponsored by the IUSSP in conjunction with the conference, provided a venue for feminist activists associated with US foundations to lobby for explicit recognition of women's rights.[22] The Final Plan of Action did add several paragraphs, which in general terms proclaimed the need to raise women's status and articulated their right to education, health care, and inclusion in development on an equal basis, justified as human rights.[23] This formal support for women's rights was juxtaposed to paragraphs, heavily lobbied for by the Holy See, that reaffirmed the family as the basic unit of society and as both the locus of decision making and the object of demographic analysis. Thus the Bucharest conference documents archive the unresolved tensions of the Malthusian couple in the contraceptive age, which subordinated women's autonomy to the social interest in stable (hetero-patriarchal) families. To the extent that they supported this muddled approach to women's individual rights, postcolonial nations also aligned with the hegemonic masculinity of demographic convictions.

Initially Notestein and his colleagues expressed great disappointment with the Bucharest conference, noting that it was "a humbling experience for everyone."[24] John D. Rockefeller III's speech at the conference was a significant source of that disappointment. In it he distanced himself from the population control prioritization of aggregate rate reduction and called for "a deep and probing reappraisal of all that has been done in the field of population." In particular, the speech contained a long section written by Adrienne Germain that positioned women as citizens, not as objects to be acted upon, arguing that family planning programs had to address the social and legal discrimination women faced in order to be effective.[25] The speech was well received by conference attendees from developing nations. To longtime Population Council staff it was a betrayal.[26] Their disappointment increased the very next day when the conference delegates formally rejected the US-backed proposal to make small families an international norm.[27] Council staff were even more distressed following the conference, when Rockefeller sought to refocus the council's mission.[28] However, its commitment to research remained, and with it the grounds for continuing dominance of demographic quantification practices in the creation of population knowledge.

Despite setbacks at the conference, the demographic establishment found much to work with in the Plan of Action. As one postconference analysis stated, the plan contained "a great deal of good sense" but it was "buried under the clichés of every ideological point of view."[29] The language that had bridged differences at the conference drew on the metaphors the council had used for years regarding responsible parenthood and customary morality. The right to have access to reliable scientific information and contraceptives was also affirmed. Moreover, the council saw opportunity in the leeway the Plan of Action gave to governments to develop their own programs and the plan's endorsement of international NGO cooperation on research.[30] While resenting the dissipation of urgency, demographers had retained authority over population knowledge. Thus, after 1974 research focused more attention on the impact of development programs on population, as evidenced by the Population Council's establishment of the *Population and Development Review* in 1975. In addition, raising the status of women as a rationale for contraceptive programs and fertility rate reduction received some endorsement in demographic circles. In part this shift reflected the growing influence of the reproductive rights movement, which sought to reorient the priorities of the field to center on women's needs and rights.[31] But, at the same time, Joan Dunlop, whom council demographers blamed for Rockefeller's apostasy, noted with concern, "Too many short-term opportunists from the population field are jumping on the women's bandwagon . . . I am beginning to fear that the demographic imperative may do to the 'status of women' what it did to family planning."[32] By this she meant that the status of women might be reduced as a means to secure demographic ends, as family planning had been, in her view.

This concern was justified. The status of women was already a sticky object in the discourses of modernity. As shown in chapter 5, the demographic transition located nations within its stages based on how well societies managed the excess reproductivity of the "as-if" woman of natural fertility. That metric did not change; it still defined the population problem. In the aftermath of the 1974 conference, demographers configured a new rate gap to justify the urgent need for population control through family planning. That statistical measure, *unmet need,* defined as the gap between those said to want no more children and those practicing contraception, appropriated and recalibrated the oft-cited lack of access that concerned birth control advocates. Following from Notestein's earlier observation that it was best for family planning programs to start with those already motivated to limit fertility, council demographers demonstrated by ever more sophisticated calculations that most of the world supported the concept of fertility control and wanted small(ish) families but did not have the contraceptive means to fulfill that desire.

To build their case, they drew on knowledge, attitudes, and practice (KAP) surveys, which asked women their desired family size, pregnancy histories, and contraceptive practices. An important element of the public relations efforts to persuade governments to establish programs, KAP studies were initially intended to assure reluctant government officials of the readiness of their people for family planning. After Bucharest, KAP studies were used to demonstrate that family planning programs were not meeting the demand for contraceptives. The calculation of unmet need is complicated. A demographically precise figure would include only those not using contraception who were at risk of becoming pregnant. For example, women who are not using contraception because they were already pregnant do not need contraception and should therefore be excluded. But women move in and out of that category over the course of any year. The literature in the 1970s and 1980s includes discussion of how best to approximate that fluctuation in aggregate risk. The standard measure, however, also builds on linguistic elision. Unmet need was generally stated in the literature as "those who say they want no more children but are not practicing contraception." However, in calculating that figure, demographers included all non-users whose number of children was already equal to or greater than their desired number.[33] This was done retrospectively to compensate for surveys that did not ask explicitly if respondents wanted to limit future fertility. Yet this methodology means that need was judged by demographers' inference, not individual declaration. Again at the level of technical detail, the counting procedures amplified the urgency of the situation. With mathematical precision and inferential sleight of hand, surveys of unmet need became a vital tool with which to advance the population control agenda in the 1970s and 1980s. With it, demographers argued that significant fertility reduction could be accomplished simply by meeting that existing unmet need for contraceptives.[34] Thus, with another muddling of the aggregate for the individual, family planning–based population control programs could be represented not as imposing on others, but as simply meeting individual desires. It is but one more example of demographic quantification that continued to use women as a means to an end.[35]

The 1994 International Conference on Population and Development

Feminist activists struggled continuously after Bucharest to shift population control priorities both in local programs and in international discourse to make women the subjects, not the objects, of policy. One focus of these struggles was to contest the conceptualization of unmet need to include those excluded from services as well as the restricted options and poor service.[36] It would take another twenty years for transnational feminists to succeed in reinserting their perspective on women's needs and rights as the proper center of fertility control policy and practice. That network grew as sites of resistance to demographic population control came together and built global alliances in the emerging global civil society that developed through the 1975–1985 UN Decade of Women.

As preparations for the 1994 International Conference on Population and Development (ICPD) in Cairo got under way in 1992, a transnational activist coalition organized through the International Women's Health Coalition (IWHC) at the instigation of Latin American reproductive health activists, built a powerful network, Women's Voices '94, to coordinate feminist efforts to shape the Cairo conference Programme of Action.[37] Joan Dunlop and Adrienne Germain, who had struggled with the gendered geopolitics of population of US foundations, led the IWHC. The majority of participants in the alliance it anchored were feminists and health activists of the Global South, such as Gita Sen of India, Sandra Kabir of Bangladesh, Peggy Antrobus of Barbados, Jacqueline Pitanguy and Sônia Correa of Brazil, Amparo Claro of Chile, Claudia Garcia-Moreno of Mexico, Marie Aimee Hélie-Lucas of the Women Living under Muslim Law International Solidarity Network, Florence Manguyu of Kenya, and Bene Madunagu of Nigeria. Together they developed and distributed the Women's Declaration on Population Policies, which laid out the ethical principles, minimum program requirements, and necessary conditions for the development of "population policies that are responsive to women's needs and rights."[38] Initially signed by representatives of twenty-five groups, the IWHC circulated the Draft Declaration to over 100 women's organizations, which gave input on its final language. When it was finalized, the Draft Declaration again circulated widely and accumulated over 2,200 individual signatures before the conference.[39]

Seeking to state a positive goal for the feminist reproductive rights agenda, the principles enunciated in the Draft Declaration linked women-centered analyses of reproductive health, human rights, and sustainable development, which has since come to be called the reproductive health paradigm. Reflecting the influence of activists from the Global South, the Draft Declaration rejects the demographic convictions about poverty and population. It decries the abstract focus on aggregate statistics, contending that, divorced from concrete social circumstances, they elide the impact of the "unequal distribution of material and social resources" in which women conduct their reproductive lives.[40] The transnational feminist alliance explicitly challenged the underlying demographic assumption that "individual welfare would be advanced" by efforts to "assist, persuade, or induce individuals to increase or to decease their fertility to meet socially desirable goals."[41] Programs built on that assumption, it concluded, subordinate women's individual reproductive self-determination to the social goal of fertility decline and thereby reverse the appropriate balance of social and individual welfare. Instead, it asserted, "Women's empowerment is legitimate and important in its own right, not merely as a means to address population issues."[42] It condemned the use of economic incentives as coercive, especially when the value of the incentive is out of proportion to the value of other resources available to women. Finally, the document brought the effects of mortality on fertility patterns into focus and urged greater attention in population policies to reducing infant, child, and maternal mortality.

The impact of the political skill of transnational feminist activism on the Cairo conference is clear in the resulting Programme of Action. Gender equity, the empowerment of women, and the elimination of violence and discrimination against women were identified as the "cornerstones of population and development-related programmes." The goal of population policy, the Programme of Action declares, is to promote reproductive health, which it defines as "the capability to reproduce and the freedom to decide if, when, and how often to do so."[43] For the first time in a UN population document, this reproductive right is specifically identified as a woman's right.[44] It gives mild endorsement to the notion of gender equity in sexual and social relationships.[45] It also repeatedly calls on nations to develop policies against sexual violence, exploitation, and trafficking of women and children.[46] The explicitly demographic sections of the Programme of Action are brief. Incentives, disincentives, demographic targets, and quotas are disparaged, along with all forms of overt coercion.[47] The quantitative goals are stated in terms of population stabilization and reduction of the proportion of unmet contraceptive need. With these guidelines, feminists attempted to shift the focus of statistics from macro-level rates to micro-level assessment of how well women's needs are met.[48]

In assessing the Cairo conference, demographers have lamented the turn away from concrete demographic targets for reducing population growth.[49] Reasserting their epistemic authority, demographers suggest that the shift in emphasis produced by feminist activists signals a turn away from sound policy.[50] Three intertwined concerns were raised in the demographic literature that resonate with longstanding demographic convictions about sound population knowledge and policy. Demographers objected to the relative weight given in the Programme of Action to the components of population policy. Policies focused on the empowerment of individual women rather than on population trends were said to obscure societal-level population needs and goals.[51] Although women's empowerment and health might be good goals, they were not the appropriate focus for population policy because individual empowerment and health could, at best, have only an indirect influence on aggregate population dynamics.[52] Policy focused on the individual would be especially inadequate, they argued, in regions "where there is little demand for contraception, where economic development is stagnant or declining, and where religion and culture tend toward pronatalism."[53]

Second, as the above quotation suggests, demographers reasserted the neo-Malthusian truism: "Rapid population growth limits a nation's ability to improve the standard of living of its people." Furthermore, "policy that ignores population as an aggregate phenomenon and that regards fertility decline as secondary to other objectives may be less likely to aid the development process."[54] The demographic momentum of past growth alone could overwhelm development efforts if not appropriately managed. Demographers reiterated that improving human conditions required measures that directly affect fertility decisions and thus growth. Moreover, voluntary measures now would eliminate the need for more drastic measures later.[55]

Finally, demographic commentators argued, subordination of demographic projects to feminist goals in the Cairo Programme of Action reflected what can happen when policy is too strongly influenced by politics. In assessing the conference, demographers noted that "the formulation of policy is a quintessentially political process; as such it is almost inevitably accompanied by some mobilization of bias. In most policymaking situations," however, "the political process is to some extent informed by serious scientific analysis, which acts as a counterbalance to ideological and political preferences."[56] Feminist activists, they observed, in "returning to the earliest roots of the birth control movement"—to anarchism and radical feminism—wielded substantial influence over the ICPD recommendations regarding development and women's rights. The result, the commentators noted, was that the chapter on demographic goals "was severely cut and lost much of its punch."[57] They determined that in this case, "serious scientific analysis" of population issues was outweighed by postcolonial and feminist ideology and politics. For demographers, the successes of transnational feminist efforts in Cairo were "divisive," providing fresh evidence of the "important function demographic science . . . serve[s] as a corrective for the excesses of policy advocacy in population matters."[58] Demographers, still committed to population control, recommended a return to demographic convictions as a sound empirical basis for effective policy.[59]

The initial demographic assessments of Cairo overstate the extent to which demographic convictions were displaced by feminist ones. The Programme of Action did not ignore aggregate statistics. While it eschewed numeric targets, it included figures for life expectancy, birth rates, total fertility, growth, unmet need, and so on, in its description of the state of the world population. Although the Programme of Action did not support demographers' action agenda, the mid-twentieth-century demographic narrative still reigns in the population facts. Furthermore, the demographic narrative—unprecedented growth, the imbalance of demographic and economic processes, the need for demographic transition—structured the rationale and objectives of the population section of the plan. Thus the objective is stated as follows: "Recognizing that the ultimate goal is the improvement of the quality of life of present and future generations, the objective is to facilitate the demographic transition as soon as possible in countries where there is an imbalance between demographic rates and social, economic and environmental goals, while fully respecting human rights. This process will contribute to the stabilization of the world population, and, together with changes in unsustainable patterns of production and consumption, to sustainable development and economic growth."[60] The document assumes that every country should and eventually would go through "the demographic transition,"[61] again extending the authority of the midcentury demographic narrative of population change.

The feminist alliance did not reject the demographic narrative of unprecedented and problematic growth or the idea that nations need to go through the demographic transition.[62] Their primary concern was that women's needs be recognized in the process.[63] Likewise, they embraced the concept of unmet need; they simply wanted to recast it to reflect women's lived experience. In this way, the transnational feminist health coalition did align partially with demographic convictions.[64] Furthermore, the demographic narrative of unprecedented growth that must be contained remained one of the pillars of the document. The demographic narrative of population change was positioned along with development, women's empowerment, and environmental sustainability as the grounds for population and development policy. Demography retained its authority over population facts, and within them Malthusian anxieties about the future.

Toward More Sound Population Knowledge

The science-politics distinction that grounds demographic epistemic authority has also led historians to misinterpret the field's influence in population politics. For instance, although Matthew Connelly notes that "the idea of a 'population crisis' provided the catalyst," he nonetheless characterizes the population control establishment as "a system without a brain."[65] He describes the United Nations as designing censuses and population projections, the Ford Foundation as the research financier, the Pathfinder Fund and IPPF as starting clinics and lobbying officials, the Population Council as the developer of new contraceptives and trainers of new experts, and "well-connected individuals" and "elite groups" as those "who worked the corridors of power to claim a place at the top of the international agenda." But he overlooks the effects of the circulation of demographers and demographic data in and through these sites.[66] As the foregoing analysis has shown, the demographic social world as nurtured and sustained by the OPR and the Population Council formed the epistemic center of the population establishment. The configuration of knowledge, power, and affect promulgated in it had profound effects on social scientific, political, and popular understandings of modern population dynamics and their control.

By excavating the elements brought together by the statistical measurements developed and deployed in mid-twentieth-century demographic social worlds, this analysis illuminates how the numbers worked to normalize fertility control as a cultural competency within both the hegemonic masculine narrative of modern marriage and the modernization paradigm that subordinated women and ranked men and nations on scales of economic and cultural difference. By investigating the configurations of the demographic social worlds and inferential practices, it elucidates the political, ideological, and affective commitments that organized both. The demographic claim that its commitment to mathematics distinguishes it from population politics is itself a gendered political claim. And the persuasiveness of midcentury population facts hinges on familiar tropes of gendered coloniality delivered in the flat monotone of positivist science. The gendered coloniality of demography is not just a gloss on the data, but is formative to demographic procedures of reasoning about fertility and population dynamics. Those procedures were assembled in the space created by credibility contests with the birth control movement and anticolonial nationalisms. Thus, demography was a party to the controversy of reproductive politics after all.

Moreover, despite the claim to impartiality and the monotone voice of positivist science by which the numbers were conveyed, sentiment was not absent from demographic figures. Demographic sentiments are not merely a gloss on the data, but are foundational to demographic epistemology. The familiar Malthusian affect—anxiety about the excess fertility of others—triggered the configuration of the binary of natural and controlled fertility. Neo-Malthusian fear was and is instantiated in demographic figures of natural fertility—the "as-if" women who make no effort at contraception across their reproductive life span, regardless of the consequences for society. As we have seen, however, they were the product of the aggregation methodology, not nature. That aggregation methodology decontextualized fertility from lives as they were lived and lost. On the one hand, sex was and is irrelevant to demographic fertility models. Calibrated to the "natural" rhythm of the biological model of the ovulation cycle, the demographic metric disconnects risk from the situated gendered and geopolitical circumstances, interpersonal relationships, structures of feeling, and fleshy worlds of women's reproductive lives.[67]

Demographic transition theory built on the assumptions of natural/controlled fertility, coupled with cultural differentialism, constructed a progressive model of social evolution. The imaginary European past thus grounded a single universal narrative of human civilization at the top of which resides the neo-Malthusian man. Observing the feminine figures of fertility at a distance, he understands that the future depends on managing their excesses, and that he is best prepared to do so. The most prolific and downtrodden of the midcentury "as-if" feminine fertility figures were the rural village women of Asia, represented as subject to traditional patriarchy and endless childbearing. The reproductive events attributed to these objectified figures were (and are) the gauge with which to assess each nation's position on the scale of modernity. They were and are the actionable objects with which to shape the size and character of human populations. Midcentury demographers were not concerned to rescue them but to manage the danger they represented. At the same time, discussion of gender inequity was limited to that already recognized and redressed in the West, which the Global South was to emulate. As the case study of the demographic configuration of India's population explosion shows, the neo-Malthusian narrative occludes the effects of colonialism and mortality. Exploitative economic relationships between the developed and the developing world are illegible in the transition's evolutionary stages. Moreover, although growth rates hinge on the statistical relationship of fertility and mortality rates, mortality is barely registered in the charts that projected unprecedented increase in human populations and its negative economic consequences. Only fertility mattered.

The masculinity narrative of reproduction assembled in twentieth-century demography, therefore, was not simply about authorizing contraception. It resecured hegemonic masculinity by closing down discursive space in which gender-differentiated negotiations of heteronormative sexuality and marriage might have occurred. And, at the same time, by framing the population crisis and world poverty entirely as a problem of others' differences and deficits, it narrowed the space in which to renegotiate the ranks of masculine mastery. If we are to hold international birth control advocates and governments accountable for coercive population control programs, as we should, then we must also fully recognize the epistemic and affective convictions of the population knowledge that hailed them with both the panic of numbers and the promise of family planning.

Feminist demographers and health activists have taken on the sexism and ethnocentrism of demographic knowledge, deploying ethnographic methods to reassess the interpersonal, structural, and institutional dynamics of reproduction. Following the tenets of transnational feminist theory, much of this feminist work focuses on local reproductive practices.[68] In so doing, it endeavors to resituate the multiple and diverse bargains and strategies women pursue in managing the precariousness of lived intimacy and the mortality of bodies in specific social-historical locations. This work is invaluable for building strong advocacy networks to continue the difficult work of articulating the terms of reproductive justice. Yet, it is not clear that feminist scholarship has had a significant impact within the intertextual fabric and the social worlds of population knowledge. Post-Cairo quantitative requirements have once again depoliticized reproductive health and stymied social change.[69] They reassert universal narratives that marginalize the multiple, diverse, and historically contingent circumstances in which reproductive lives are lived and lost. Thus feminist scholars and activists must continue to navigate a terrain shaped by a troubled relationship to demography and its population facts.

In my view, three intertwined approaches are called for by these circumstances. Feminists must continue to challenge the reach of demographic narratives and practices. As the analysis herein has shown, bringing a transnational feminist lens to bear enables the decolonization of population knowledge by identifying the ideological and affective economies organizing that knowledge. Such reflexive intellectual work will continually be needed to illuminate the many varied experiences of difference and support epistemologies emerging from them. It is also needed to resist the call to rescue figurative downtrodden women so often implicit in feminist accounts as well as the familiar pull of Malthusian anxiety about the impact of incompetent and irresponsible "as-if" others.

But feminists must also engage aggregate population figures at the level of their calculation. As this analysis has shown, we cannot simply trust the numbers. They are haunted by the gendered geopolitics, affects, and epistemic commitments of midcentury hegemonic masculinity. Bringing a transnational feminist lens to bear on the figures, it is possible to excavate how the numbers work, both the sophistication of the mathematical equations and the political commitments of the inferential technologies that organize them. The tools of textual analysis can elucidate what was enumerated, how it was enumerated, and what inferential logics support statistical analyses of them. In that regard, the contours of gender differences and hierarchies that organize demographic figures and the population imaginary offer a rich site for further analysis. The goal of such analyses is not to throw out statistical accounts of the world, but to illuminate the moments/sites of ideological and affective attachments that produce mistaken accounts.[70] Interrogating the technical details and the host of inferential steps that produce demographic figures creates the space to imagine population otherwise, to reconfigure our understanding of aggregate population trends and to delimit the proper scope of statistical forms of knowledge. It can thereby serve as the base for building a new trust in numbers, one based on an understanding of the specific complex histories of their making.

Finally, feminists must challenge demographic narratives that habitually position women-centered analyses on the side of politics and disembodied demographic figures on the side of science. This distinction was built into the midcentury credibility contests between demographers and birth control advocates. It relies on gendered associations between science and quantification to insulate demographic knowledge from political scrutiny and facilitates the dismissal of feminist, postcolonial, and related marginalized knowledges as unreliable. Thus feminist scrutiny of demographic figures must also continually dismantle claims that impartiality is produced by quantification itself. In the emerging era of big data, it behooves those marginalized by the gendered geopolitics of quantification to recall that demography is and always will be political mathematics, and thus commit to build population knowledge with inferential practices derived from our own situated epistemologies.

Notes

Chapter 1. Matters of Vital Importance

Epigraphs: Haraway 1997, 218; Porter 1995, 213.

1 Ehrlich 1968. The back cover emphasizes Ehrlich's scientific credentials as a noted professor of biology.

2 Ibid., 15–16 (emphasis in original). The press of the Indian populace on Western visitors is a recurring theme in Western literature and memoirs. See Cohn 1996, 10; Robert J. C. Young 1995, 97–98. On the contemporary period, see Hartmann 1995, 103, on J. D. Rockefeller III's memoir. Greene 1999, 164, and Schoen 2005, 216–217, also quote Ehrlich's opening vignette.

3 Ehrlich 1968, 17. Retreating to the safety of numbers in the face of the press of India's strangeness is a gesture with a long history. See Cohn 1996, 8; Appadurai 1996.

4 Ehrlich 1968, 18.

5 Ibid., 22. Labels for regions of the world are always imprecise and politically inflected. In sections where I am referring to the historical episteme of demography, I use the heavily freighted terms *Third World, First World,* and *undeveloped, developing,* or *developed* because these terms reflect the nomenclature of the time and the objects configured by demographic knowledge. When speaking from the perspective of the present, I use the geographic terminology of *Global North* and *Global South.* Although these terms do not adequately convey the political configuration of the world, they are the best approximate terms currently in use.

6 Ehrlich 1968, xi, defines population control as "the conscious regulation of the numbers of human beings to meet the needs, not just of individual families, but of society as a whole."

7 On Moore, see Critchlow 1999, 4–5, 16–18, 20–33. Critchlow 1999, 32, notes that Moore had distributed 1.5 million copies of his pamphlet by 1968. Ads for Moore's pamphlet appeared regularly; see, for example, the *New York Times,* October 11, 1959, E6. The 1970 edition of Ehrlich's book indicates

that the book, a best seller, had twelve printings in less than two years. On the modern social imaginary, see Taylor 2004.

8 On the historiography, see for example Linda Gordon 1976, 393–394; 1990, 398–399, and 2002, 284–285; Critchlow 1999, 4–5, 55–56, 150, 153–157; Connelly 2006, 89–90, 162, 165, 186, 199. Connelly, 1–4, in particular, begins with a standard recitation of the demographic narrative of unprecedented growth.

9 The term *population explosion* is often credited to Kingsley Davis 1945, 1. See Connelly 2006, 309. Frank Notestein 1945a, 98, also used the explosion metaphor, but he was referring to Europe, whereas Davis was referring to the Global South.

10 See Latham 2000, 1–8, 17, 21–22, on modernization as a response to postwar threats and the confidence of modernization theorists in science and their own expertise.

11 Borrowing from physics, demographers define population dynamics as the interactive processes of fertility, mortality, and migration that shape the size, distribution, and composition of population. See Lorimer and Osborn 1934 for an early theorization.

12 Notestein 1950a, 186.

13 On the influence of American demography, see Greenhalgh 1996, 30; Riedmann 1993, 102–106; Symonds and Carder 1973, 37; Connelly 2008, 11.

14 See Latour 1987, 215–241, on the importance of "centers of calculation" to "made science."

15 See Wilmoth and Ball 1992 on press coverage of population issues.

16 The population establishment is generally defined as the network of medical, policy, family planning, and demographic organizations and actors concerned with controlling global population growth through contraception. Of the vast literature on population control, see Critchlow 1999; Connelly 2008; Greene 1999; Bashford 2014; Reed 1983. For feminist critiques, see Schoen 2005; Linda Gordon 2002; Ahluwalia 2008; Briggs 2002; Hodges 2008; Murphy 2012; Gita Sen, Germain, and Chen 1994. For critiques of demography, see Greenhalgh 2008, 2005; Hartmann 1995; Bandarage 1997. On the reproductive life sciences, see Clarke 1999.

17 *Coloniality* refers to the Eurocentric epistemology of the social sciences, which, using race, configures a hierarchy of nations that positions the North Atlantic world at the pinnacle or center and the rest below and behind it. See Mignolo 2002, 62, 75–80, 82–86; Quijano 2007, 168–171, 176.

18 Connelly 2008, 122–123, notes that the demographic transition model was a description that became prescriptive. But he does not question its credibility as a description of population change.

19 Smith 1990, 78–79, notes that the conditions of the production of facts are not visible within them. Curtis 2001, 30–33, 308, observes that invisibility of the constructedness of population is testament to its status as made science. See also Smith 1990, 69, on methodological configurations of fact;

she writes, "A fact . . . coordinates the activities of members of a discourse, a bureaucracy . . . profession. . . . It coordinates the activities of anyone who is positioned to read and has mastered the interpretive procedures it intends or relies on."

20 As Kingsley Davis 1951b, 4, declares, "So far as possible the mechanics of this analysis, though fundamental, is kept from intruding. . . . The emphasis is on the results and conclusions rather than the techniques of the analysis."

21 Greenhalgh 2005, 357.

22 On articulation of social and epistemic genealogy see Porter 1995; Somers 1996; Poovey 1998; Schweber 2007, 8.

23 See Somers 1996, 68, on the context of discovery as a focus of inquiry into historical epistemologies and Gieryn 1995, 405–407, on the inscription of local circumstances of its production into scientific knowledge.

24 Clarke 1998, 15–17, defines social worlds as communities of mutual practice and discourse. See also Gieryn 1995, 412–413.

25 Porter 1995, 114, observes that calculation is "always an interaction between quantitative methods and administrative routines." Smith 1987, 29, defines authority as "a form of power" that has the "capacity to get things done in words." On intertextuality and intertextual hierarchies see Hooper 2001, 122–125; Smith 2006, 66–67, 77–87.

26 Haraway 1997, 218; Foucault 1979, 194; Desrosières 1990; Mignolo 2002, 69. Desrosières 1990, 200, defines things that hold together as things "which display qualities of generality and permanence" and "transcend the contingency of particular cases and circumstances."

27 Somers 1996, 63, defines knowledge culture as a "broad spectrum of contested truths," problematizations, and knowledge justification strategies "delimited" by historically specified "ways of thinking and reasoning." It defines "the boundaries of what is conceivable in our historical imagination."

28 Hacking 1992, 143.

29 See Ahmed 2004b, 11ff, on affect and action.

30 See Rotman 2000 on mathematics as a symbolic system and Curtis 2001, Greenhalgh 2008, and Schweber 2007, 19, on the cultural work of numbers.

31 Ahmed 2004a, 121; 2004b, 11, 92. See also 2004a, 119–123, 127–130; 2004b, 89–93, 194–195.

32 See Avery Gordon 2008, 18, 198–201, on structure of feeling; on imagining and learning lessons otherwise, see Avery Gordon 2008, 22–25; Spivak 1992, 775; McCann and Kim 2013, 477–478.

33 On hegemonic masculinity see Connell 2005a, 2005b; Carrigan, Connell, and Lee 1985; Connell and Messerschmidt 2005; Hooper 2001; Demetriou 2001; Gardiner 2002.

34 Connell 2005b, xviii, 71–72; 2005a, 72. See also Hooper 2001, 55–59.

35 Connell and Messerschmidt 2005, 832; Connell 2005b, 77; Carrigan, Connell, and Lee 1985, 592.

36 Connell 2005b, 72; Gardiner 2002, 14; Connell and Messerschmidt 2005, 843–844, 846, 853.
37 Connell and Messerschmidt 2005, 846; Curtis 2001, 37.
38 Connell and Messerschmidt 2005, 842.
39 Hooper 2001, 74.
40 On emphasized femininity see Connell 1987, 183–188; Connell and Messerschmidt 2005, 848; Hooper 2001, 54. Connell and Messerschmidt 2005, 837, note that hegemonic masculinity theory has been criticized as heteronormative.
41 Connell 2005a, 73, 75.
42 McClintock 1995, 13. See also Mignolo 2002; Quijano 2007; Cooper and Stoler 1997, 33–37.
43 Malthus 1989 [1798].
44 On haunting, see Somers 1996, 63–66; Avery Gordon 2008, xvii–xviii.
45 Cooper 2003, 162.
46 Haraway 1997, 218.
47 "Women's Declaration" 1994, 33.
48 *Reproductive performance* is defined by demographers as biological capability modulated by social institutions, especially family form, religion, and an Orientalist notion of cultural heritage. In contrast, the estimate of natural fertility is divorced from social relations. See Szreter 1996, 26–27; chapter 4 of this volume.
49 Ahmed 2004b, 11, 92. Objects are fetishized in her view when they become stuck. Immobilized, they no longer circulate, but intensify the affective value of the associations to which/by which they are stuck.
50 Escobar 1995, 3, 21, 4; Latham 2000, 2–8. On the discovery of world poverty see Escobar 1995, 21–39; O'Connor 2001, 99–102, 113–117.
51 In fact, the term *Third World* was coined by the French demographer Alfred Sauvy in 1952 to describe those nations caught between the two sides of the Cold War—the First World of the West and the Second World of the Communist nations. At the 1955 Bandung Conference, perhaps the first international postcolonial conference, the term *Third World* was taken up by nonaligned nations forming a new power bloc that became the Group of 77. See Prashad 2007, 6–15, 31–50; Robert J. C. Young 2001, 157, 51, 59, 191–192; Greene 1999, 72–74. On the social sciences and the modernization regime see Latham 2000, chapter 2.
52 Escobar 1995, 45.
53 On normalizing judgments in population statistics, see Curtis 2001, 4, 311–313; Foucault 2007, 56–63.
54 Mignolo 2002, 66–69, 70–74.
55 Balibar and Wallerstein 1991, 57, 21. See also Mignolo 2002, 78–80.
56 Balibar and Wallerstein 1991, 55, 18, 17. See also Stoler 1995, 91; Cooper and Stoler 1997, 10; Goldberg 1993, 42, 64, 77–84; O'Connor 2001, 99. Stern 2005 notes that the policy achievements of American eugenics—immigra-

tion restrictions and compulsory sterilization—remained in place until the social movements of the 1960s.

57 Hooper 2001, 41–48, defines masculinism as power and privilege staked in hegemonic masculinities. On the crisis of masculinity of this period, see Hooper 2001, 66–69. On Parson's sex roles theory see Carrigan, Connell, and Lee 1985, 553–557. On Parson's functionalist sociology in demography, see Riedmann 1993, 104; Calhoun and VanAntwerpen 2007, 406; Szreter 1993, 673, 684.

58 O'Connor 2001, 106, 107–117, and Yuval-Davis 1997, 62–64, discuss gender, family forms, and modernization. See also Stern 2005, chapter 5, and Ladd-Taylor 2001 on the 1950s shift in eugenics from biologizing racial differences to biologizing gender differences.

59 Somers 1996, 68–70.

60 Symonds and Carder 1973, 78–79. See also Dudley Kirk, "World Population Problems and American Foreign Relations," March 7, 1956, Ad Hoc Committee Meeting, and meeting minutes, RAC-PC, IV3B42, B1, F7.

61 Kingsley Davis 1945, 1, the first to use the term *demographic transition,* opens his account with bomb imagery.

62 Stoler 1995, 8, 50, 108, defines cultural competencies as bourgeois cultivated habits and self-disciplines of personal hygiene, comportment, and language use by which the European bourgeoisie distinguished themselves from diverse colonial others. See also 12, 45–53, 90, 205–206.

63 Hacking 1990.

64 Hooper 2001, 59.

65 See chapter 5. Although through far different modalities and with different implications, those demographic figures set the standards of reproductive self-mastery for white women and women of color in the Global North. See Collins 1999, who observes that judgments abroad come home to shape reproductive control of US women of color. See also Yuval-Davis 1997, 60–61, Mani 1990, and Hodges 2008, 136, on the position of women as a marker of modernity.

66 For historiography on contests between population controllers, Communists, and the Catholic Church in international arenas, see, e.g., Connelly 2008; Critchlow 1999; Sharpless 1995.

67 Yuval-Davis 1997, 30–35.

68 Existing sources on mid-twentieth-century demography are overwhelmingly insider accounts. See Szreter 1993; Riley and McCarthy 2003; Hodgson 1983, 1988, 1991; Demeny 1988; Greenhalgh 1996; Presser 1997. On the rhetoric of moderation as a hallmark of twentieth-century sciences, see Clarke 1998, 254. Historiography of modernization and development has overlooked the importance of demographic figures to that history. Escobar 1995 is one exception.

69 Demography has never been fully institutionalized as an independent discipline. Its more tentative standing may well result from its close as-

sociation with reproductive matters, from which even the most rigorous mathematics cannot redeem it. See Clarke 1998, 233–258, who notes that the reproductive life sciences suffered from association with the "taint of human sexual interaction" (91). Greenhalgh 1996, 30, notes that fertility control was the largest component of demography after World War II.

70 Van de Tak 1991, 6; Hodgson 1983, 1988, 1991; Demeny 1988, 555–557; Riley and McCarthy 2003, 39–41. Even Thomas Malthus laid claim to mathematics as an insulator from politics. See Porter 1986, 26. The effort to insulate itself from politics is also identified as the cause of the relative theoretical poverty of the field, which identifies the demographic transition as its chief contribution to social theory. Later in life, Frank Notestein concluded that "the action elements of the Association were probably less anxious to capture us than we were to avoid capture" (quoted in Van de Tak 1991, 6).

71 Szreter, Sholkamy, and Dharmalingam 2004, 8, 9–11. Greenhalgh 2005 and 2008, which analyze the math panic in the governance of China's population, are exceptions.

72 Gieryn 1999, 14. See also Gieryn 1995, 419; 1999, 10; Curtis 2001, 31.

73 Curtis 2001; Porter 1986, 1995; Hacking 1982, 1986, 1990, 1992. Appadurai 1996 and Cooper 2003 discuss population numbers in colonial contexts. Greenhalgh 1996, 33–36, also notes the gendered quality of demographic mathematics. Similarly, Clarke 1998, 272, concludes in regard to the reproductive life sciences that "scientific frames permeate lay frame[work]s and . . . the power to define has been and remains largely masculine."

74 Curtis 2001, 26–27, 33–34. If anything, there is a solid belief that censuses undercount the true number.

75 Elsewhere, drawing on the work of Rhacel Parreñas, I define social processes as places where bodies, ideologies, and institutions converge to be shaped and to shape each other. See McCann and Kim 2010, xi, 148, 184–202.

76 Porter 1995.

77 Campbell 2000, 36.

78 Gieryn 1995, 419; see also 1999, 10, 14.

79 Butler 1998, 3. She continues, "In this sense, then, the subject is constituted through the force of exclusion and abjection, one which produces a constitutive outside to the subject, an abjected outside, which is, after all, 'inside' the subject as its own founding repudiation." We will see this played out in chapter 3.

80 Irene Taeuber as quoted in Lunde 1981, 483. Of course, the gentlemen were overwhelmingly white Westerners. Taeuber's status as a published demographer is often used by demographers as evidence that their field was not men-only. She worked as a research associate at OPR from 1936 to 1973. See Van de Tak 1991, 7.

81 Greenhalgh 1996, 33.

82 The acknowledgments section of midcentury demographic texts often in-

cluded long lists of the names of such individuals and/or expressions of appreciation to wives and women staff for the preparation of charts, graphs, and manuscripts. See, for instance, Coale and Hoover 1958, Kingsley Davis 1951b, and Berelson et al. 1966. Irene Taeuber, cited above, was a demographic wife. She was married to Conrad Taeuber, a statistician for the US Census.

83 See Mignolo 2002, 70–73, on the double bind in which such scholars are caught: their work is either so similar to Eurocentric scholarship that it is dismissed as not making a significant contribution, or it is so different that its credibility is doubted. Feminist scholars face a similar double bind.

84 "The Ad Hoc Policy Committee: Proposed Meetings on the Philosophy and Public Relations of the Rational Control of Family Size," January 24, 1955, 1, RAC-PC, IV3B42, B1, F3. See also Kirk 1956, "Population Problems," 4, RAC-PC, IV3B42, B1, F7.

85 Connelly 2008, 230, which likewise concludes that "India was the cutting edge of the population control movement."

86 See Mignolo 2002, 71, and 2007, 159, on border thinking as a tool for decolonizing knowledge.

87 Disciplinary histories emphasize demographers' "early" rejection of eugenic biodeterminism as proof of demography's liberalism. See Ramsden 2008, 393; 2002, 861. Ramsden's important work (2008, 2003, and 2002) explores many of the ties between the two with regard to domestic US population matters. However, some may find his conclusion that social eugenics supported greater democratization and human freedom to be overly generous. Bashford 2014, 240, 13, 342, 343, argues that the connections between them are quite close because Malthusianism constitutes a "formative" logic of eugenics. Linda Gordon 2002, 280–282, also notes the close relationship of eugenics and demography, calling for its careful study.

88 There has been a great deal of debate about what eugenics was and is. I follow Stern 2005, 11, who defines eugenics as "better breeding" practices. This encompasses both the class and race components of biological determinism most commonly associated with eugenics, but it also opens analytical space in which to explore the influence of eugenics on gender regimes.

89 For feminist critiques of demography, see Greenhalgh 1995, 1996, 2005, 2008; Riley and McCarthy 2003; Susan Cott Watkins 1993; Presser 1997.

90 Greenhalgh 1995, 24. On feminist critiques of social sciences, see Stacy and Thorne 1985; Smith 1987, 1990.

91 Greenhalgh 1995, 25, and Riley and McCarthy 2003, 50–51, 77–80, also note that demography is a Eurocentric, modernist discipline.

92 Riley and McCarthy 2003, 115. These arguments have been important for incorporating feminist analyses into demography. There is a good deal of compatibility between my analysis and that of Riley and McCarthy 2003 and Greenhalgh 2005. But borrowing Latour's 1987 distinction, their work

focuses rather more on the science-making processes, whereas I focus more on the cultural work of the made science.

93 All rely on Dennis Hodgson's work as the standard account of demography's history.

94 Curtis 2001, 34, notes that there are no independent grounds for "estimating the accuracy" of population figures because they are all produced by the same protocols.

95 For the historiography of statistics see Curtis 2001; Porter 1986, 1995; Appadurai 1996; Hacking 1986, 1990, 1992; Poovey 1998; Cooper 2003; Schweber, 2007.

96 Following the work of Donna Haraway and Banu Subramaniam, I conceptualize natural and cultural objects such as reproductive bodies as co-configurations. I follow their convention of marking the rejection of a nature/culture divide by using the compound single word. See Haraway 2000, 106, and Subramaniam 2014.

97 Rotman 2000.

98 There is a rich literature on gender and nation. See Hooper 2001, Yuval-Davis 1997, and Sinha 2004. See Yuval-Davis 1997, 3–4, on categorization.

99 Hacking 1992, 140.

100 Porter 1995, 213.

101 See Moya 2001, 477–478, on this feature of postpositivist realism.

102 Kirk 1996, 363–364, credits Carr-Saunders 1936 with articulation of the concept of the "small family system." Kirk also says that Notestein relied on Carr-Saunders's "compilation of data" and "discussion of demographic processes" in his initial formulation of demographic transition theory. See Carr-Saunders 1936, 107–116, 243–259, on the origin and problem of the small family system. See also Bashford 2014, 234.

103 Yet, such codes are never fully in the conscious control of authors, and their meanings always exceed authorial intent. See Hooper 2001, 122, on this point.

Chapter 2. Rereading Malthus

Epigraph: Curtis 2001, 24.

1 On demographers' claims to Malthus, see Hauser and Duncan 1959 and chapter 5.

2 For a recent critique, see Amartya Sen 1994.

3 Poovey 1998, 280, 283.

4 Winch 1987, 5–10.

5 McClintock 1995, 4–8. See also Mignolo 2002, 68–69, 75–85; Quijano 2007.

6 Foucault 1991, 2007.

7 Malthus 1989 [1798], 10, 15, 11–12.

8 Ibid., 13, 15.

9 Ibid., 11, 14, 13. On misery and vice, see 17–23.

10 Moral restraint as a positive check appeared first in the 1803 edition. See Winch 1987, 36–37.

11 Malthus 1989 [1798], 19, 16, 10, 16–17.

12 Ibid., 11, 17.

13 McClintock 1995, 45; Malthus 1989 [1798], 301. See also McClintock 1995, 39–45, on this feature in Darwin's conceptualization of the family of man.

14 Malthus 1989 [1798], 17–19. Malthus, 32, describes American Indian women as libertine. See Petchesky 1985, 38, 41–42, on Malthus's assumption of patriarchal control of women and Davenport 1995, 420, on Malthus's concern with the dangers of illicit sex.

15 Pateman 1988, 23, 50–54, 87–88. See also Yuval-Davis 1997, 12–13, 78–79.

16 Godwin 1820, quoted in Petchesky 1985, 40.

17 Foucault 1978, 103–105, also takes a masculinist perspective, identifying the Malthusian couple as the "privileged object of knowledge" through which modern discourses of sexuality normalized and regulated procreative practice. However, as Stoler 1995, 4, notes, for him sexuality rather than reproduction links the life of the individual and the species.

18 See James 1979; Winch 1987, 13.

19 Malthus 1989 [1798], 24.

20 See also McClintock 1995, 38–42, on temporal distancing in Darwin's writing. Darwin credits Malthus's population principle with providing the pressure he thought was necessary for selection to operate. Winch 1987, 7; Bashford 2014, 232.

21 Malthus 1989 [1798], v, 25, vi.

22 Mignolo 2002, 67–68.

23 McClintock 1995, 32, 36–39.

24 Malthus 1989 [1798], 252, 266, 254–255. Book 2 discusses Norway, Sweden, Russia, France, England, Scotland, and Ireland. France required two chapters as well.

25 On the cultural work performed by such texts, see Cooper 2003.

26 Malthus 1989 [1798], 25.

27 Ibid., 27.

28 Ibid., 23, 27–29, 32–34. Davenport 1995, 422–423, also notes the degradation of women in the *Essay*.

29 Malthus 1989 [1798], 116, citing Sir William Jones's *Works,* vol. 3. On famine as an effect of colonialism in India, see chapter 6.

30 Malthus 1989 [1798], 124, 128, 125, 132. On this point, Malthus 1989 [1798], 126, cites Jesuit missionary reports: "It cannot be said in China, as it is in Europe, that the poor are idle, and might gain subsistence if they would work. The labors and efforts of these poor people are beyond conception." This is an example of the colonial other being used to admonish the metropolitan other.

31 Stoler 1995, 8. See also 12, 45–53, 90, 133–134, 183–184, 205–206.

32 Ibid., 19.

33 With regard to colonialism, Malthus 1989 [1798], 13, notes that morality prevents the extermination of indigenous inhabitants, but concludes that "the process of improving their minds and directing their industry would necessarily be slow" and "of variable and uncertain success."
34 Ibid., 20–21. He specifically laments that accurate marriage and mortality rates are not available for disaggregated population subgroups, especially the lower classes.
35 Foucault 1978, 140, further notes that these instrumentalities "mark[ed] the beginning of an era of 'biopower.'" See also Foucault 2007, 1991. Foucault's observations on population and governmentality have generated a wealth of scholarship. However, it has tended to overlook gender politics. On statistical history, see Porter 1986, 1995; Appadurai 1996; Hacking 1986, 1992, 1990; Curtis 2001; Poovey 1998; Cooper 2003; Desrosières 1990; Schweber 2007. On governmentality, see especially Burchell, Gordon, and Miller 1991; Rose and Miller 1992; Rose 1991, 2007; and Curtis 2001, 42–44, who offers an insightful critique of Rose and Miller.
36 Curtis 2001, 8.
37 Foucault 2007, 21.
38 Hacking 1986, 222; Foucault 1978, 25. See also Hacking 1982.
39 Foucault 1978, 25; see also Foucault 2007, 66–79; Hacking 1990.
40 Curtis 2001, 24.
41 Desrosières 1990, 200. See also Poovey 1993 on the relationship between numbers and narrative in statistics.
42 Appadurai 1996, 123, 117; Curtis 2001, 308.
43 Foucault 2007, 42, 378.
44 Curtis 2001, 41.
45 Foucault 2007, 75.
46 See Horn 1994; Cohn 1996, 1987; Poovey 1995; Davenport 1995.
47 Foucault 2007, 75.
48 Curtis 2001, 24. See also Cooper 2003 on economic statistics.
49 Anderson 1991, 166; Curtis 2001, 26–27.
50 Curtis 2001, 24. Censuses also comprise temporal fictions. They purport to be a count of the same people at specific intervals over time. I follow Curtis's use, 34, of the term *census-making* instead of *census-taking* to highlight its constructedness.
51 Ibid., 24. See also Desrosières 1990, 197–200.
52 Curtis 2001, 31–33. Curtis 2001, 17, notes that census-making depends on "mobiliz[ing] consistent observational protocols" and "coherent observational practices at the local level."
53 Ibid., 18.
54 Schweber 2007, 20. See also Curtis 2001, 311–312.
55 Appadurai 1996, 120; Curtis 2001, 5, 12; Mignolo 2002; Quijano 2007.
56 On normalizing judgment see Foucault 2007, 56–63; Curtis 2001, 4, 311–313.

57 Appadurai 1996, 121.

58 Ibid., 125. See Kingsley Davis 1951b, 163, on the precise instructions on caste designation given to Indian census workers.

59 Appadurai 1996, 120, 125; Curtis 2001, 32–33, 310.

60 See Chandrasekhar 1959, 39, who makes this point as well. Without a birth certificate it is not easy to decide on the age of an individual. Birth registration is generally worse than death registration, especially when it occurs outside of institutionalized medical establishments. This is because death registration is often necessary for funereal purposes. See United Nations 1952; Alborn 1999.

61 Curtis 2001, 24–25.

62 Cohn 1996, 8; 1987; Curtis 2001, 37–38. See also Appadurai 1996; Anderson 1991; Nobles 2000; Omi 1997; Andersen 1988; Haraway 1997.

63 Curtis 2001, 35, 308.

64 See Curtis 2001, 25–26, 37–38, and Hacking 1986 on genealogy and future, and Stoler 1995, 133–134, 183–184, on essence.

65 On Quetelet's average man, see Porter 1986, 52–54, 102–103, 107–109; Hacking 1990, 105–114; 1992, 147–148; Schweber 2007, 62; Curtis 2001, 19–21.

66 Surveys became a major form of social knowledge-building in the mid-twentieth century. On the average man in mid-twentieth-century survey statistics, see Igo 2007.

67 Porter 1986, 103.

68 On the law of error, see Porter 1986, 6–7, 43–44, 95. Hacking 1992, 141–149, notes that there is nothing natural about transferring it to social statistics.

69 Porter 1986, 104. See also Hacking 1990, 2.

70 See Butler 1998, 3, on abject bodies and beings.

71 Pearson is recognized as having mathematized statistics by pioneering applications of the normal curve to frequency distributions of discrete events. See Porter 1986, 128–129, 296–314; 2004; Kevles 1985, 31–33, 37; McCann 1994, chapter 4; Petchesky 1985, 72–73, 84–89; Ramsden 2002.

72 On Spencer's prominence in the United States, see Breslau 2007, 46–47.

73 Szreter 1996, 73. Pearson appears to have made similar proclamations after numerous national censuses. Szreter says Pearson's calculation was initially based on 1895 data from Copenhagen. See also McCann 1994, 107.

74 Francis Walker first raised the issue based on results of the 1870 census. See McCann 1994, 109–110. See also Bashford 2014, 246–251.

75 Notestein 1982, 657, makes this observation in a section titled "The Fallout from Public Health."

76 Depopulation theory was the principal theoretical perspective of the field in the 1930s, when low birth rates in the United States and Europe dropped the net reproduction rate below replacement levels. See Notestein 1981, 486; McCann 1994, 186; Carr-Saunders 1936, 243–246.

77 Homi Bhabha as quoted in Escobar 1995, 9.

78 Yuval-Davis 1997, 11.
79 See Briggs 2002 on Puerto Rican women represented as submissive and as coupled to hypermasculine men. See Collins 1999 on images of black women's reproductivity, and May 1999 and Britt 2000 on images of childless (white) women in the United States.
80 Curtis 2001, 31–35.
81 They are vital in the double sense of being both lively and pertinent. *Vital statistics* is the name given to statistics about birth, death, marriage, disease, and health. On their history see Schweber 2007.
82 Hacking 1990, 3–5.
83 Curtis 2001, 42. See also 39–42, where he cogently argues that Foucault falls into a realist perspective on population knowledge when he refers to populations in terms of demographic figures.
84 Michel Foucault as quoted by Stoler 1995, 84. See also Foucault 2007, 11–12.
85 Curtis 2001, 35, 310. See also Escobar 1995 on this within the development regime.
86 Hacking 1986, 228, 229, 234. See also Yuval-Davis 1997, on gender, nominal identities, and nationalism.
87 On moral panics, see Hall et al. 1978.
88 Curtis 2001, 24, 28. See also Hacking 1992, 141, 154.
89 Porter 1986, 103. Porter 1986, 31, notes that Quetelet and Malthus were acquainted.
90 Porter 1995, 101–106.
91 Foucault 2007, 60–61.
92 Calculation of natural increase excludes the effects of migration. That is what makes it natural. See Dublin and Lotka 1925, 305. Although the article is authored by Louis Dublin and Alfred Lotka, Lotka is generally credited with the mathematical formulas. See also Ramsden 2002, 866, and 2008, 393, 396–398, on Lotka's equation, Raymond Pearl's influence on it, and Ansley Coale's refinements to it.
93 Taeuber 1946, 256.
94 Kingsley Davis 1951b, 87.
95 Dublin and Lotka 1925, 306–307, 319.
96 These tools were developed by actuaries to help accurately price life insurance. They calibrate the probability of death in one-year increments of age. Estimates of life expectancy at birth are derived from life tables. See Porter 1995, 101–113. See chapter 6 for the use of life tables in population projections.
97 Notestein 1943, 169, declares that "much of the history of past events may be read from the [age/sex] pyramids."
98 Taeuber 1946, 257.
99 Kingsley Davis 1951b, 87.
100 Doubling time relies on the generational time factor of stable population theory. See Notestein 1944, 427, 429; 1948, 252; 1950b, 91; 1951, 100–101;

1952, 21–23; 1969b, 37; Kingsley Davis 1951b, 88; 1956, 317; 1957, 15.

101 Dublin and Lotka 1925, 307–309. Susan Cott Watkins 1993, 570, n. 2, notes that a two-sex model greatly complicates the mathematics.

102 Dublin and Lotka 1925, 311; Susan Cott Watkins 1993, 560; see also chapter 4.

103 In 1933, Warren Thompson and Pascal Whelpton coauthored the "first systematic attempt" to consider the social and economic implications of each of the components of US population growth, birth, death, and migration. See Notestein 1982, 654. See also Appadurai 1996, 120.

104 Lorimer 1951, 31. See also chapter 6.

105 Curtis 2001, 29.

106 More recently, as Greenhalgh 2008 has shown, demographic data authorized a panic of numbers that underwrote the one-child policy in China. See also Connelly 2008 and Hartmann 1995. See Chatterjee and Riley 2001 on representations of happy families in Indian population control posters. On the promise of happiness, see Ahmed 2007/2008 and 2010.

107 See Hooper 2001, 5, 64–65, on this figure in the pages of the *Economist*.

Chapter 3. Narratives of Exclusion, Mechanisms of Inclusion

Epigraph: Radway, in Gordon 2008, viii.

1 Lunde 1981, 479; Hodgson 1991, 9. Cf. Bashford 2007, 2008, and 2014, which situate migration as the central element of population knowledge in this period. See also Fairchild 1934 on the field's research agenda.

2 Lorimer 1981, 488; Szreter 1996, 14. The Scripps foundation was set up to investigate population problems in Asia.

3 Ramsden 2002, 863.

4 On Burch see Linda Gordon 1990, 394; 2002, 281.

5 See Coale 1983, 4; Ramsden 2002, 869; 2003, 548; Connelly 2008, 84.

6 In fact, I was struck by how often the story was recounted in disciplinary histories when I began research for this book. Lunde's paper, one of four on a panel, was entitled "The PAA at Age 50." The others all referred to the founding episode in the same tone and tenor as Lunde did. See Lorimer 1981; Kiser 1981; Notestein 1981. Kiser 1981 provides a list of the previous accounts of the founding meeting. It includes a 1934 conference presentation, a tenth-anniversary conference paper, both by Henry Pratt Fairchild, and a 1953 presentation by Kiser himself. Frank Lorimer, Frank Notestein, and Clyde Kiser also provided histories at a 1971 conference commemorating the Milbank Memorial Fund. See Notestein 1971; Lorimer 1971; Kiser 1971. See also Ryder 1984, 7.

7 More recent histories include Hodgson 1991, 19–21, and Greenhalgh 1996, 32, 34–36. Connelly 2008, 52, 84, 106, discusses the exclusion of Sanger and Osborn's role in the field. Ramsden 2003, 562, Riley and McCarthy 2003, 64, 71, and Presser 1997, 299, briefly mention the story. Perhaps the most recent version is the 2008 PAA Conference Panel, "How Did We Get Here

and Where Are We Going?" See http://geography.sdsu.edu/Research/Projects/PAA/paa.html. Accessed July 26, 2015.

8 Lunde 1981, 481, read the archived reminiscence of Frederick Osborn. All other accounts agree that Fairchild nominated her as vice president and that Osborn was the one who suggested that Sanger withdraw her name.

9 Lorimer 1981, 488.

10 Kiser 1953, 110, argues that the association's chief success was to give "members a sense of belonging." See also Van de Tak, 1991, 93, and Lunde 1981, 484, who note the "intimacy" of the "small group" of men who ran the early PAA.

11 Hodgson 1991; Connelly 2008; Schweber 2007. Greenhalgh 1996, 32, 34–36, and Riley and McCarthy 2003, 62, 64, 70, insightfully note the gendering of the distinction between demographic science and politics. However, they accept the disciplinary truism that politics threatened the discipline. Greenhalgh argues that it developed an overreliance on mathematics in response. I put much more emphasis on the gendering of this demarcation than they do. My observation that the narratives of the founding events are themselves instances of boundary marking is novel.

12 On social worlds of science see Clarke 1998, 15–17; Gieryn 1995, 412–413.

13 Gieryn 1999, 14–17, identifies three genres of boundary work in, 15, the "seizure or defense of epistemic authority": expulsion, expansion, and protection of autonomy. See also Clarke 1998, 269–271; Ramsden 2002, 890.

14 Clarke 1998, 15, quoting Messer-Davidow. Clarke 1998, 240, notes that in the reproductive life sciences the reputational risk of association with sex and reproduction was both a "self-constructed and self-serving myth used rhetorically" to increase its status and a fact. See Riley and McCarthy 2003, 39–41, on demographers' sensitivity regarding political influences and the "struggle to maintain scientific objectivity" as "one of the fundamental myths of contemporary demography." See Clarke 1998, 91ff; Messer-Davidow, Shumway, and Sylvan 1993, 3; and Kathy Davis 2007, 85–119, on mythmaking in origin tales.

15 Gieryn 1995, 394. See also Gieryn 1999, 4–6, 12–18.

16 Clarke 1998, 7. Feminist critiques of science have generated a vast literature. With regard to reproductive life sciences see especially Clarke 1998 and Jordanova 1989, who note that the power of definition has been overwhelmingly masculine.

17 Reed 1978, 204.

18 On Osborn's eugenics, see Ramsden 2003, 2008; Bashford 2014, 244, 258, 329, 350–51; Kevles 1985, 170–177.

19 See Szreter 1993; Hodgson 1983, 1988, 1991; Demeny 1988. On the close relationship between eugenics and demography, see Ramsden 2002, 2003, 2008; Linda Gordon 2002; Bashford 2014.

20 Lunde 1981, 479.

21 Notestein 1982, 684.

22 Clarke 1998, 25. See also 163ff, with regard to the authority of reproductive life sciences over contraception.
23 Irene Taeuber as quoted by Lunde 1981, 483. This is of course an implicit statement about both gender difference and hierarchy; demography was a white bourgeois men's club.
24 Reed 1978, 204; Connelly 2008, 84.
25 Lorimer 1981, 489; Lunde 1981, 481. On Osborn's support for birth control, see also Osborn 1937.
26 Notestein 1981, 487; Lunde 1981, 480–481.
27 Notestein 1971, 70.
28 Clarke 1998, 254. The distinction between research and propaganda was also central to Sanger's conflicts with the medical profession. See McCann 1994, 82–88.
29 Notestein 1981, 485; 1982, 682–683.
30 Smith 2005, 68, 225.
31 See Greenhalgh 1996, 31. See also Calhoun and VanAntwerpen 2007, 400–401.
32 The National Science Foundation project began in 1954. See Demeny 1988.
33 Vance 1959, 304–305.
34 Van de Tak 1991, 5–6, observes that all important decisions were made in pre-meetings of the board, thus limiting the influence of the laypersons among the general membership.
35 Gieryn 1999, x. See also Clarke 1998, 185–186, 201.
36 Gieryn 1995, 405–407, 419. See also Gieryn 1999, x, 10, 14.
37 Ramsden 2002, 858–859, notes the lack of standardized methods in 1931, saying PAA imposed "some semblance of unity and order" on the field. See also Ramsden 2003; Clarke 1998, 181; Bashford 2014, 20–23.
38 Notestein 1982, 660.
39 Gieryn 1999, xii (emphasis in original) .
40 Butler 1998, 3.
41 On boundary work, see Clarke 1998, 15–17; Gieryn 1999, 11, 14, 16–17.
42 Chesler 1992, 409; Suitters 1973, 192.
43 Gieryn 1995, 405.
44 Clarke 1998, 269.
45 Of course, the distinction between emotion and rationality, science and sentiment, is itself an exercise of gender power that subjugates women's knowledge. See Wanzo 2009, 146–147; Berlant 2008, x, 1–12, 20–22; Howard 1999; Clarke 1998, 272–273; Jordanova 1989; Gieryn 1995, 422.
46 Most versions of the story refer to Sanger only as "Mrs. Sanger," emphasizing her status as lay patron. See Lorimer 1981, 489; Notestein 1969a, 368; 1982, 660.
47 See Lengermann and Niebrugge-Brantley 1998, 11, and Ferree, Khan, and Morimoto 2007, 474, on women's precarious grasp on cultural and scientific authority.

48 Notestein 1981, 485.

49 The advisory council included only prominent scientists from the United States and Europe. However, representatives from India, China, Japan, Brazil, Argentina, and Chile did serve on the general council. See Sanger 1927, 11–12.

50 The next year, when the IUSSPP formed, Pearl was instrumental in preventing individual propagandists from dominating by limiting membership to national professional organizations. On the IUSSPP founding, see Symonds and Carder 1973, 12–15; Presser 1997, 299; 2003, 553; Connelly 2008, 69, 71; Bashford 2014, 82–85. The second *p*, for *problems*, was dropped when the group reorganized after the war in 1947. On Mallet, see Szreter 1996, 266–267, who notes that Mallet was a prominent eugenicist and a key figure in the 1911 British census analysis that produced the initial social class categories used by modern social scientists.

51 Sanger 1939, 376–377. For Pearl's role in Geneva, see Sanger 1939, 378. See also Ramsden 2002, 858; 2003, 553.

52 As Calhoun 2007, 23, notes, these firings, including those of Edward A. Ross and Thorstein Veblen from Stanford, were "pivotal events in the history of academic freedom" and development of tenure. See also Lengermann and Niebrugge 2007, 88, 93.

53 On scientism in sociology in this period, see Calhoun 2007, 28; Camic 2007, 230–232, 265–266. See also O'Connor 2001, 103; Laslett 2007, 500, on the post–World War II period.

54 Latour 1987. See also Avery Gordon 2008.

55 McCarthy and Das 1985, 29. See also Calhoun and VanAntwerpen 2007, 400–401, who note the tensions between basic and applied research in demography.

56 See Camic 2007, 230–232, 265–266; Platt 1996, 67–105.

57 See McCarthy and Das 1985, 29. Gieryn 1995, 406, notes that the chances of success in credibility contests is enhanced where the boundaries of the science are congruent with the interests of the powers that be.

58 For a detailed account of the gender exclusions in early twentieth-century sociology see Calhoun 2007, 28; Laslett 2007, 480–502; Lengermann and Niebrugge 2007, 109–110; Lengermann and Niebrugge-Brantley 1998, 11–13; Ferree, Khan, and Morimoto 2007, 474–479; Silverberg 1998, 4–5, 15–17; Breslau 2007, 60.

59 Women found few opportunities in social science departments from this period until the 1970s. If hired at all, they found themselves relegated to schools of applied and domestic science. Of course, although differently, racial segregation also limited access to higher education for people of color.

60 McCarthy and Das 1985, 27–28.

61 Laslett 2007, 500, notes the intense masculine culture of scientism in post–World War II sociology, which enshrined Parson's sex roles theory as normative gender identities and relations. See also Ferree, Khan, and

Morimoto 2007, 456–457; Silverberg 1998, 9–12; Mignolo 2002, 68–69.

62 Lengermann and Niebrugge-Brantley 1998, 11–13, make this point with regard to women sociologists: once they are excluded it is easier to caricature them in disciplinary histories.

63 Lorimer 1971, 86. See also Notestein 1982, 680.

64 Sharpless 1995, 83.

65 Connelly 2008, 169; Critchlow 1999, 180–181; Chesler 1992, 409, 300, 451.

66 Notestein 1982, 684.

67 Ibid., 679, 678. As Clarke 1998, 20, notes, the tinge of controversy was inherent in the reproductive sciences because of their intimate connection with "both the sacred and obscene."

68 Notestein 1982, 679, 680.

69 Notestein's assessment constitutes an article of faith within the field. Both the MMF and later the Rockefeller Foundation gave generously to establish demography as the scientific study of fertility patterns and their control even while they avoided funding contraceptive clinics directly. As Notestein 1981, 488, observed, the MMF "hewed the line of investigation rather than advocacy." See Gieryn, 1999, 22–23, on immediate audiences of credibility contests.

70 Clarke 1998, 91. See also 185–186, 201, 233–248; McCann 1994, 122–125; Chesler 1992, 214–217.

71 Notestein 1982, 684.

72 Ibid. 1939b, 125.

73 Thompson 1929.

74 "Science: Eugenics for Democracy," *Time Magazine* 36 (September 9, 1940): 34.

75 Notestein 1982, 659–660; 1969a, 368–369; Ramsden 2003, 558; 2008, 393; Kevles 1985, 170. In another bit of disciplinary folklore that symbolically confers scientific status upon him, Osborn is said to have insisted on taking an examination to demonstrate his proficiency at the end of his self-study.

76 Lunde 1981, 481.

77 Ibid.; Notestein 1969a; Population Council 1978, 21–25.

78 Kiser 1981, 494; Van de Tak 1991, 94.

79 Notestein 1969a, 367.

80 See Reed 1978, 205, on Osborn's social connections to Rockefeller, the MMF, and Princeton University.

81 As Connelly 2008, 106, observes astutely, Osborn was "in a better position than Sanger to manage male scientists' egos." He held a degree in English from Princeton. Sanger attended but did not complete nurses' training. Osborn was born to wealth, while Sanger grew up working class but married into wealth. These class differences accentuate the gendered differences between them. Osborn's class privilege gave him qualities associated with masculine science: education and economic independence. Acquired

through marriage, Sanger's class privileges reaffirmed her gendered position as a dependent woman.

82 Lipsitz 1988 reminds us that speaking softly and being heard is a mark of privilege.

83 Kiser 1981, 494. When founders note their presence, it is usually to reject the suggestion of racial bias that an affiliation with eugenics implies. See Lorimer 1981; Notestein 1971; 1969a. In the 1990s, disciplinary histories acknowledged and discounted the field's historical engagement with eugenics, following the by now reflexive distancing narrative. See McNicoll 1992; Hodgson 1983, 1988, 1991. Ramsden 2008, 392–393, and Bashford 2014, 23, 240–246, 343, also argue that demographers discount the field's continuity and overlap with eugenics. See also Ramsden 2003, 550, 562ff, for his critique of Hodgson 1991 on this point.

84 Osborn served on the board from 1930 to 1970. The board was a virtual who's who of leaders in the field, including Kingsley Davis, Otis Dudley Duncan, Dudley Kirk, Clyde Kiser, Frank Notestein, Norm Ryder, and Pascal Whelpton.

85 See Hodgson 1991, 25, n. 7, for the text of the 1937 PAA charter. This language remained into the 1970s.

86 In a similar vein, Clarke, 1998, 112, argues that eugenics and genetics "extend[ed] the study of development . . . from the individual to the population." See also Haraway 1997; Bashford 2014, 13, 23, 240–246, 341.

87 Notestein 1939b, 126; Stix and Notestein 1940, 155.

88 Hodgson 1991, 19, on Dublin and Thompson. Theodore Roosevelt expressed concern in 1905 that Anglo-Saxons were committing race suicide because women were shirking their duty to the state to contribute suitable progeny for the next generation of citizenry. See McCann 1994, 110.

89 Ramsden 2003, 558–562, and 2002, 859, also notes that eugenicists and demographers were united by statistics and by common nonscience competitors.

90 Many eugenicists held important academic positions. Henry Pratt Fairchild, a sociologist, held a professorship at New York University. Raymond Pearl held a professorship at the Johns Hopkins School of Hygiene and Public Health.

91 Following the work of Haraway and Subramaniam, I conceptualize natural and cultural objects such as reproductive bodies as co-configurations. I follow their convention of marking the rejection of a nature/culture divide by using the closed compound. See Haraway 2000, 106; Subramaniam 2014.

92 See Ramsden 2002, who argues that descriptions of this debate in recent demographic histories have served to symbolize the distance from eugenics that demographers believe they established in the 1930s.

93 Pearl's research is discussed at length in chapter 4. See also Ramsden 2002, 885–887. Reed 1978, 413, n. 21, observes that in his 1974 interview with Notestein, Notestein claimed to have encouraged Pearl to pursue the study

in the hope that the results would change his mind. As Ramsden comments, Pearl's conversion also serves to ratify the scientific integrity of the field because he was persuaded by the data to change his earlier position.

94 Ramsden 2002, 869–878; 2008, 393; Kirk 1960, 305.

95 Notestein 1982, 684.

96 Ramsden 2008, 394–395, and Stern 2005, 152–153, note that the shift away from biologism by Osborn and others in the 1930s focused the attention of eugenicists on population control to manage heredity. See also Stern 2005, 150–181. My analysis owes much to her excellent analysis of gender and eugenics.

97 Ramsden 2003, 561. Ramsden labels Osborn a social eugenicist and argues that social eugenics was more liberal than previous versions. While I concur that Osborn's eugenics was based more in culture than in genetics, I do not agree that this changed the political inflection substantially. For insider histories on Osborn, see Lorimer 1981, 488; Notestein, 1969a. Osborn's moderation is also used as evidence that after World War II demography disengaged from eugenics, and thus racism. See Hodgson 1983, 1988, 1991; McNicoll 1992, 401.

98 Osborn never abandoned his commitment to eugenic management of genetic endowment. He simply believed the time was not yet ripe for such policies. In private correspondence, he argued that it was necessary to practice "crypto-eugenics" because the public was not ready to accept a transparently eugenic agenda. See Connelly 2008, 163. At the end of his career, Osborn facilitated a rapprochement between population geneticists and demographers and guided the American Eugenics Society to a new identity as the Society for Social Biology. His last book, published in 1968, makes an early case for the development of sociobiology. See Ramsden 2003, 560; Ramsden 2008; Osborn 1968.

99 Ramsden 2008, 394, also makes this point.

100 See Reed 1978, 413, on its constructed objectivity. See Ramsden 2008, 393; 2003, 558–561, on the influence of Frank Lorimer and Harry L. Shapiro on Osborn's thinking and his commitment to placing eugenic philosophy on solid scientific footing.

101 Notestein 1971, 73, identifies *Dynamics* as "clearly one of the most important demographic works of our generation." See also Ramsden 2003, 560; Kiser 1981, 494; Notestein 1969a, 367.

102 Osborn published a revised edition of *Preface to Eugenics* in 1951. See also Osborn 1958, 13–15, 51–68. Chapter 5 discusses the Population Council in greater depth.

103 Lorimer and Osborn 1934, 347; Osborn 1940, 285–291.

104 Lorimer and Osborn 1934, 156, 126, 344, 252; see also 126–141; Osborn 1940, x, 42–44.

105 Osborn 1940, 119, 96, 95, 78. Osborn opposed eugenic policies based on group identity because they included too many of the wrong people and

missed too many of the right people. See Taeuber 1946, 260, which praises Osborn's approach to quality in *Preface*.

106 Osborn 1940, 66. At that point, effective eugenic management of the genetic heritage of the nation would be possible. On the relative weight of heredity and environment, see also 17–19, 41, 56, 59–60, 67, 73, 96, 101, 136–137, 198; Ramsden 2003, 560.

107 Balibar and Wallerstein 1991, 18, 20, 57.

108 See Lorimer and Osborn 1934, 21, 179, 36, 54, 140–141, 325–326; Osborn 1940, 121.

109 Osborn 1940, 41, 94–98.

110 Ibid., 96; Lorimer and Osborn 1934, 343–344.

111 Osborn 1940, 148. See also 133; Lorimer and Osborn 1934, 184. Use of the term *retardation* to describe cultures speaks to the continued conflation of individual and aggregate, of psychology and culture, by demographers and their eugenic compatriots of the period. On the conflation of individual and aggregate in eugenics and genetics see Haraway 1997 and Clarke 1998, 112. On the use of this language in relation to emerging nations of the Global South see chapters 5 and 6.

112 Osborn 1940, 136, 337. See also Lorimer and Osborn 1934, 252, 340, 343–344. They follow Pearson's class-based logic of racial degeneration.

113 Lorimer and Osborn 1934, 347.

114 Osborn 1940, 71, 148.

115 Lorimer and Osborn 1934, 185 (emphasis in original). In this case, they are speaking specifically about poor rural whites in the Appalachian region of the nation. See also Osborn 1940, 191; he states that for those "at a close margin between income and a decent living, children are a chief cause of poverty."

116 Balibar and Wallerstein 1991, 21.

117 Lorimer and Osborn 1934, 337; Osborn 1940, 78.

118 Osborn 1940, 119, 98. See also 42, 73, 148.

119 Balibar and Wallerstein 1991, 18.

120 Osborn 1940, 119.

121 Lorimer and Osborn 1934, 252.

122 Osborn 1940, 118–119.

123 Ibid., 109.

124 Ibid., 194.

125 Ibid., 109. See Lorimer and Osborn 1934, 329, who describe rural areas as locations of cultural backwardness.

126 See Osborn 1940, 38, 71–75, 78, on the need for "further scientific studies of genetics capacities of different races"; see also 285–291. Osborn 1940, x, notes that "questions of cultural quality will come first," but "inevitably the attempt to influence births will include an effort to cast out the worst of these hereditary factors and to continue the normal or superior." See also Lorimer and Osborn 1934, 343, 347.

127 Balibar and Wallerstein 1991, 22–24.
128 Lorimer and Osborn 1934, 99. For additional examples of this gesture, see Osborn 1940, 43, 78.
129 See Ahmed 2004b and 2007/2008 on affective economies and race.
130 Osborn 1940, 143. The "Emperor-God" is a highly charged reference to Japan on the eve of the US entrance into World War II.
131 Moreover, like other eugenicists, he did not go so far as to advocate repeal of the earlier generation's primary policy successes: immigration restriction based on ethnicity and compulsory sterilization laws. See Stern 2005, 154–155. Throughout the fifties, Osborn cautioned against immigration as a possible solution to population pressures in developing nations, as discussed in greater depth in chapter 5. With regard to sterilization, Osborn 1940, 19, 30–35, suggested that there would be reason for greater reliance on it as methods improved for identifying carriers of genetic defects. See also Lorimer and Osborn 1934, 334–335, and Fairchild 1934; Bashford 2014, 330–339.
132 Szreter 1996, 73.
133 See Balibar and Wallerstein 1991, 24.
134 Ibid., 49 (emphasis in original). See also 56, where Balibar notes that theoretical racism, of which cultural differentialism is one, "necessarily has built into it sexual schemes."
135 Stern 2005, 155. Stern's insights refer to marriage counseling, but are applicable to population science as well. See also O'Connor 2001, 107–113, on the gendered ground of theories of the culture of poverty, which is a variation of cultural differentialist racial discourse.
136 Osborn 1940, 212–213. See also 240.
137 Stern 2005, 150–181. Balibar and Wallerstein 1991, 100–105, likewise note the importance of the nuclear family to racialized nationalism, observing, 100, that it is "ubiquitous in the discourse of race." Again, their analysis ignores gender.
138 Lorimer and Osborn 1934, 338–339.
139 Camic 2007, 278–279; Stern, 2005, 152–156.
140 Lorimer and Osborn, 1934, 334. Osborn 1940, 156, observes that appropriate clinical practices would be based on "less publicity," greater medical expertise, and "a more serious attempt to introduce contraception among the people who need it."
141 Osborn 1940, 109, 149, 143, 198, 104. For his version of the proto-demographic transition see 103–105. See Carr-Saunders 1936, 106–116, on the origin of the small family system.
142 Osborn 1940, 143, 133, 156, 265, 193.
143 On cultural competency see Stoler 1995, 8, 12; chapter 2.
144 Osborn 1940, 119, 157ff, 195, 198, 232–252. Osborn 1937, 389, argues that the low birth rates of the urban superior classes needed to be "restored to normal condition."

145 Lorimer and Osborn 1934, 332–333. They, 332, define the demographically normal family as one with three or more children. For Osborn, responsible parenthood would always include the obligation of the "best elements" to have as many children as they could support and educate. See Osborn to Dr. Walter Chivers, October 21, 1957, and Osborn to William Vogt, November 8, 1957, RAC-PC, IV3B42, B25, F369.

146 Clarke 1998, 267, cites demography as an example of disciplinary "formation and coalescence" that occurred "only when fed, nurtured, and legitimatized by extrascientific interests and social worlds." For a full discussion of the institutionalization of demography see chapter 5.

147 See also Symonds and Carder 1973, 82–84, and Suitters 1973, 79, 83, on the IUSSP and the United Nations.

148 Porter 1995, 213.

149 Chesler 1992, 426. Osborn letter to Mrs. G. J. Watumull, May 24, 1958, RAC-PC, IV3B42, B19, F305.

150 Likewise, Frank Notestein lent his name to the early meetings, but did not attend. See Chesler 1992, 408–410, 419, 574, on gender tensions.

151 Chesler 1992, 425–426. The two women were Irene Taeuber and Dorothy Swaine. In 1965, Mary Bunting became the first female board member. See Population Council 1978, 157.

152 See Ramsden 2001, 4, which concludes that with the council's establishment, the founders believed that "the realm of fertility control" was "placed firmly in the hands of science." Concern that association with birth control would tarnish their scientific credentials continued throughout this period. See also chapter 5; Reed 1978, 287; Ferguson interview, 1974, 28–29. Ferguson's interviewer, James Reed, framed the conversation in terms of the dominant distinction between the PPFA and the Population Council when he, 29, says, "Well, you were a propaganda organization." See also Connelly 2008, 156–160, on the articulation of the goal of containing population growth in the Global South at the founding meeting of the Population Council.

153 See "Conference on Population Problems, June 20–22, 1952," 9, RAC-Office of the Messrs., Rockefeller records, series O, B1, F4.

154 Notestein 1981, 488. See also Osborn to William Vogt, president, PPFA, January 20, 1958, RAC-PC, IV3B42, B25, F369, and Osborn to Robert Lentz, National Council of Churches, October 18, 1955, RAC-PC, IV3B42, B1, F5.

155 Osborn to Mrs. Phillip Pillsbury, January 16, 1958, RAC-PC, IV3B42, B19, F305. See also Ramsden 2001, 4; Connelly 2008, 156–157; Reed 1978, 287; Clarke 1998, 226; Ferguson interview, 1974, 28. In the Population Council documents see also Osborn to Dorothy Brush, January 25, 1954; Dudley Kirk to William Vogt, April 30, 1954, RAC-PC, IV3B42, B24, F367.

156 Osborn to Margaret Sanger, October 16, 1935, MSP-LC, reel 74, and Francis Bangs to Mrs. Thomas Lamont, March 15, 1934, MSP-LC, ABCL file, reel 24.

157 Ad Hoc Committee Meeting Minutes, May 11, 1955, 3, RAC-PC, IV3B42,

B1, F4 (emphasis in original). PPFA president William Vogt was explicitly excluded.

158 Osborn to Rufus Day, June 9, 1957, RAC-PC, IV3B42, B19, F305; Osborn to C. P. Blacker, June 3, 1957, RAC-PC, IV3B42, B19, F306.

159 Ferguson interview, 1974, 42–44.

160 See chapter 4; Ramsden 2002, 886; Clarke 1998.

161 Osborn to Vogt, January 20, 1958, RAC-PC, IV3B42, B25, F369, and Paul Henshaw, director of research, PPFA, to Frederick Osborn, October 30, 1953, RAC-PC, IV3B42, B24, F367. See Kirk to Osborn and files, April 9, 1956, RAC-PC, IV3B42, B24, F368, noting that the PPFA might be "an appropriate agency" for "laboratory analysis" and clinical testing of contraceptives. Kirk's memo responded to that from Osborn, March 20, 1956, proposing that the council allocate funds to set up a standards and evaluation bureau in the PPFA. See also DK to FO and files, April 17, 1956, RAC-PC, IV3B42, B1, F7.

162 See Bill Esty, PPFA interim director of research, to Population Council, April 23, 1954, RAC-PC, IV3B42, B24, F367. See also PPFA Research Program Brochure, "The Control of Human Reproduction," March 1957, 7, RAC-PC, IV3B42, B25, F369. See also the exchange of memos, Osborn to Kirk, March 20, 1956, and Kirk to Osborn and files, April 9, 1956, RAC-PC, IV3B42, B24, F368. Still focused on basic reproductive science, in the end the council absorbed and revitalized the National Committee on Maternal Health (NCMH) as its biomedical research division. The NCMH, which had sponsored biomedical research on contraceptive agents in Europe and the United States in the 1930s, provided the basic research and testing for the modern IUD under council auspices. A male-dominated medical contraceptive research group, it languished after the death of its founder, Robert Latou Dickinson, until taken over by the council. See Takeshita 2011, 14–15, 17, 38–41, 58; Clarke 1998, 185–190.

163 Sanger was denied entry in 1948. Notestein was in Japan as part of the Rockefeller Foundation–funded population survey team. See "Enclosure to Dispatch No. 669 dated October 12, 1948 and Memorandum of Conversation, October 5, 1948," RAC-PC, IV3B42, B1, F3. Sanger had longstanding ties to Japan, and was well regarded. When she visited in 1954 she became the first US woman to address the Japanese Diet, and in 1959, the government awarded her the Third Order of the Sacred Crown. See PPFA, *News Exchange,* March 1950, 1, 5, MSP-SSC, B68 F1, on denial of entry; and Suitters 1973, 120–124, 313, on that and later visits and honors.

164 Ruth Regan (Osborn's secretary) to Drs. Nelson and Notestein, October 18, 1956; Osborn to Drs. Nelson, Notestein, and files, October 15, 1956; Frederick Osborn to Mrs. Charles Bosanquet, September 27, 1956; Osborn to Vera Houghton, February 19, 1957, and March 7, 1957; DK to FO and files, June 25, 1957, RAC-PC, IV3B42, B19, F305. Osborn also enlisted others in this effort; see Robert Cook to Dr. Henry Hadow, to Vera Houghton, and to Osborn, March 15, 1957, RAC-PC, IV3B42, B19, F305.

165 Connelly 2008, 160. The council also gave financial support to the Population Reference Bureau. Osborn declined to become a public sponsor of the Population Reference Bureau, saying he wanted to avoid taking on too many commitments. He did maintain an informal relationship with the bureau. See FO to DK and files, May 6, 1958; Osborn to Robert Cook, December 7, 1953, RAC-PC, IV3B42, B25, F381; October 7, November 4, and December 10, 1954, RAC-PC, IV3B42, B25, F384.

166 Chesler 1992, 426, 451, 427, 407, 410, 419; Osborn to Kirk and file, May 6, 1958, RAC-PC, IV3B42, B25, F381; Osborn to Sanger, June 25, 1957; Sanger to Osborn, June 14, 1957; Osborn, interoffice memo, June 25, 1957; Osborn to Mr. Rufus Day, June 9, 1957; Osborn to Blacker, September 15, 1958, RAC-PC, IV3B42, B19, F305; Osborn to Sanger, April 3, 1957, RAC-PC, IV3B42, B19, F306.

167 Connelly 2008, 206, indicates that Osborn and Blacker met at the 1927 conference in Geneva. See Osborn to Blacker, October 8, 1958, which responds to Blacker's letter to Osborn, October 1, 1958, RAC-PC, IV3B42, B19, F306. See also Osborn to Blacker, September 15, 1958, and November 21, 1958; Blacker to Osborn, November 5, 1958, RAC-PC, IV3B42, B19, F305.

168 Suitters 1973, 197.

169 IPPF, "Special Opinion Poll," August 28, 1958, RAC-PC, IV3B42, B19, F305. The conversation between Blacker and Osborn was spurred by this poll.

170 Suitters 1973, 192, 235. Elsie Ottesen-Jensen, a longtime Swedish sex education and family planning activist, was elected as IPPF president. Blacker served as administrative chairman, managing activities at the headquarters in London.

171 See Chesler 1992, 409, 300, 451, on demographers' relationship to the early IPPF. On Sanger's faith in women, see 409. On doubts about the health focus of family planning, see Symonds and Carder, 1973, 37–38; "Social Aspects of Family Planning and Fertility Control" draft program, Second Arden House Colloquium on Human Fertility, January 22–24, 1954, RAC-PC, IV3B42, B24, F367. Connelly 2008, 283–285, also cites disparaging comments from the male demographers at the Ford Foundation about the all-volunteer staff and loosely affiliated local associations that comprised the IPPF in the 1960s.

172 Osborn to Blacker, November 21, 1958, RAC-PC, IV3B42, B19, F305.

173 Throughout her memoir, Suitters comments on the ongoing tension between the goals of improving maternal and infant health of individual women and aggregate population control. The shift in her language from the beginning to the end reflects the growing dominance of the population science perspective. See Suitters 1973, 170–171, 198, 201, 235. See also Frances Hand Ferguson interview, 1974; Ferguson, 17, likewise comments that the "overpopulation bit" did not come in until the "mid-fifties." She attributes the change to the entrance of men into leadership of the birth control movement.

174 Smith 2005, 68, 225, notes that institutionalization is never still; there is always motion demanding a refixing of boundaries.

175 See Van de Tak 1991, 716–718. See also Sanger to Henry Pratt Fairchild, January 28, 1932, MSP-LC, reel 68. Using the rhetoric of the founding meeting, she declined Fairchild's invitation, saying there would be "more harmony if laymen and propagandists are few in your organization." She declined again in 1935. See Sanger to Frank Lorimer, January 31, 1935, MSP-LC, reel 68.

176 Chesler 1992, 287, 410–412. Chesler notes that Sanger felt the slight to her expertise by male scientists strongly. Sanger certainly saw herself as more than a celebrated figure, a "great woman of her time."

177 Chesler 1992, 221–222, comments on her complicated personality, using the gendered binary of reason and emotion. She also notes the harshness of David Kennedy's characterization of Sanger's need for recognition, saying, 16, that his biography is "marred by his unexplained animus and patronizing attitude towards his subject." Although the connection is indirect, I would note that Kennedy's account was written at the height of the Population Council's credibility on population control and the proper role of birth control advocates therein. Thus it is perhaps not surprising that his work rehearses the animus and patronizing tone of midcentury demographers such as Osborn. The tone of Kennedy's account carries forward in feminist accounts. See Linda Gordon 1976, 1990, 2002; Hartmann 1995; Bandarage 1997; Ahluwalia 2008. I am not interested in defending Sanger but in illuminating the gendered politics that have influenced condemnation of her.

178 Gieryn 1999, 11–12, notes that the questions to be asked of boundary work are "*what* is it good for?" and "does anyone 'profit'?" Emphasis in original.

179 See also Linda Gordon 1976, 391, 395–398, 401–402; 1990, 386, 388–391; 2002, 279–280, 282; Hartmann 1995, 101–106, on the relationship of birth control and the population establishment. My account gives greater weight to the influence of population science on the relationship than theirs.

180 On this point I agree with Ramsden 2002, 2003, 2008.

181 Suitters 1973, 192. These statistics gained attention, as Suitters notes, at the same moment that family planning lost some of its feminism.

182 Smith 1990, 100–101.

183 Gieryn 1995, 428.

184 Clarke 1998, 272–273, also notes of the population control establishment that "the persistence and intensity of efforts in these worlds to exclude or marginalize women and women's concerns and desires regarding contraception is, today, almost shocking."

185 Connelly 2008, 297–298. See also Adrienne Germain interview, 2003, 21.

186 On the field's resistance to feminism see Greenhalgh 1996; Riley and McCarthy 2003; Susan Cott Watkins 1993; Presser 1997.

187 See Harriet Presser in Van de Tak 1991, 694–697, 708–712, 716–718; Frances Hand Ferguson interview, 1974, 46, 61–64. See also Connelly 2008, 297–298, 308.

188 I borrow this phrasing from Avery Gordon 2008, 22, who uses it to refer to sociology in general.

Chapter 4. Remaking Malthusian Couplings for the Contraceptive Age

Epigraph: Smith 1990, 4.

1 Smith 1990, 69.

2 On scientism and grand theory in sociology in this period, see Calhoun 2007, 28; Camic 2007, 230–232, 265–266, 276–278. On sociology's intense masculinism, see O'Connor 2001, 107–113; Laslett 2007, 474–479, 500–502.

3 Stacey and Thorne 1986. Susan Cott Watkins 1993, 559–560, 562, notes the absence of sex in studies of fertility.

4 Susan Cott Watkins 1993; Greenhalgh 1995; Hartmann 1995; Bandarage 1997; Presser 1997; Riley and McCarthy 2003.

5 Smith 1990, 88.

6 Ryder 1984, 8.

7 Britt 2000, 208.

8 Smith 1990, 78.

9 Berlant 2008, 5–6.

10 Haraway 2000, 106.

11 See Avery Gordon 2008, 18, 198–201. See also Ahmed 2010, 216.

12 Avery Gordon 2008, 63.

13 Smith 1990, 4, 95, 52–53. See also Susan Cott Watkins 1993, 554, whose analysis of *Demography* in 1964, 1974, 1984, and 1992 found such gender conventions in articles that focused on data accuracy, demonstrating that gender logic informed calculation practices.

14 See Foucault on "society of normalization," quoted by Stoler 1995, 84.

15 *Motherhood in Bondage* contained 470 letters, a few of which were sent by men. The appendix, 437, notes that the selected letters are but a small fraction of the 250,000 that were sent to her in the preceding years.

16 See Haraway 2000, 106. On the specific challenges to reproductive arrangements of this period, see Simmons 2009, 61.

17 Peiss 1986. See Simmons 2009, 107–111, on increased demands and opportunities for personal liberty and 71 on the binary contrast of the restrained Christian gentleman with the vigorous man of sporting culture.

18 Szreter 1996, 561.

19 Simmons 2009, 8, 12, 70, 120, 130–131; Chesler 1992, 116–121.

20 Sanger's earlier works also engaged the ideology of romantic love. See McCann 1994, chapter 2; Simmons 2009, 76–77, 94, 102.

21 Simmons 2009, 94, notes that the potential danger of daughters' independence was of particular concern. Historian Nancy Cott, quoted in Simmons

2009, 71, notes that "marriage is what 'makes the public order a gendered order.'" See also McCann 1994, 26–27, 43–46, on cultural anxiety about women's sexual autonomy.

22 Simmons 2009, 179, 94, 37, 142–143. She notes, 4, that companionate marriage was the norm by 1940. See also 104–118, 127, 72, 121–124; O'Connor 2001 107–113.

23 Simmons 2009, 138–139. See also 5–6, 37, 71, 74, 132–135, 140–143, 149–150, 167–170, 181–183, 190–194, 202, 207, on the shifting meanings of American masculinity. Simmons notes that the new model tended to ignore women's continuing economic dependence on marriage, thus offering only limited grounds for equality in marriage. On modern girl imagery globally in this period see Weinbaum et al. 2008.

24 Berlant 2008, 1–2. Simmons 2009, 119–124, in an otherwise excellent analysis of marriage reform in the early twentieth century, seems to accept the widely held assumption of that literature that the available contraceptives worked satisfactorily.

25 Sanger 1928, xiv. See also Florence 1930, 108.

26 Berlant 2008, 1–2, 5–6, 20, 12–13, 20, ix. Berlant's framework is an invaluable lens for rereading birth control movement literature as an archive of women's knowledge.

27 On the other hand, Florence 1930, 12 (see also 36), specifically eschews "sentimentality and romantic fancies." However, the personal interviews on which she relies have much of the same tone as the letters in Sanger's book.

28 Sanger 1928, xii–xiii.

29 See Berlant 2008, 12, 19–22, 27. Berlant describes sentimental realism as an ambivalent political intervention because of its underlying acceptance of conventional femininity.

30 Sanger 1928, xiii.

31 Ibid., xiv.

32 Ibid.

33 Ibid., 42.

34 Berlant 2008, 20–22. Race is not readily discernible in the letters themselves, but Sanger, xv, includes African American women in her description of the southern farm women who wrote to her. See also McCann 1994, chapter 5.

35 Sanger 1928, xviii, 3–4, also characterizes US women's dire situation through Orientalist gestures comparing it to the situation of women in India and China.

36 Sanger 1928, xi, xv, xvi, ix.

37 Karen Sánchez-Eppler, quoted in Howard 1999, 64, notes that sentimental literature "radically contracts the distance between the narrated events and the moment of their reading." Thus, readers feel as though they too are caught up in the emotions of the text. On the allusion to slavery and

the sentimental genre of women's writing, see Wanzo 2009, 146–150, who makes the case that such stories should be counted as evidence.

38 Sanger 1928, 22. I have maintained the spelling and grammar of the letters as they were published.

39 Smith 1990, 92. On numbers as distant from bodies and affects, see Latour 1987, 236–237; Appadurai 1996, 123; Hacking 1990, 1992.

40 Sanger 1928, 43.

41 Ibid.

42 Ibid., 36–37.

43 Ibid., 38. The letters also speak to women's struggle to do "their work" in the face of repeated pregnancies. Their concerns hover between anxiety about their families' suffering when they could not carry out "their work" and their own suffering when they simply carried on. Today the work they describe (washing, ironing, cleaning, canning, and sewing) would be called housework, but the letters referred only to "my work." See 177, 36–37.

44 Ibid., 50. See also 55, 57, 108.

45 Ibid., 104.

46 Ibid., 57.

47 See ibid., 16, 320, 42, for this specific wording. See also 57, 72, 78, 80, 81, 141, 162 for similar sentiments.

48 See, for example, ibid., 49, 53–54, 64–65, 71, 73, 93, 108, 118, 140–141, 154, 170, 223, 231, 233, 241, 243–244, 277, 330, 342. Women also wrote after recent miscarriages, some of which were self-induced. See 75, 77, 138, 142, 224.

49 Ibid., 141.

50 Maternal mortality rates in the United States did not decline significantly until the late 1940s, when blood transfusions and antibiotics were introduced. Leavitt 1986, 154–155, 182–190, 194; see also 170–179.

51 Sanger 1928, 256–259. Some of these letters note that both the mother and the infant died. See also letters in chapters titled "Wasted Effort," 137–154, "Voices of Children," 179–201, and "Two Generations," 202–218.

52 Marital discord was one of the central themes Sanger used to organize the selection and compilation of letters, but this theme resonates in letters throughout the book. See also Florence 1930, 101–104, 106, 119–120.

53 Sanger 1928, 20, 70.

54 For examples of this specific wording, see ibid., 37, 78, 82, 92, 98, 107, 110, 228, 238.

55 Ibid., 312, 356, 51, 81–82. On quarrels, threats, and going elsewhere as a result of the keeping-away strategy, see also 7–8, 31, 45, 51, 54, 82, 91–92, 98–99, 110, 120, 122–124, 169, 177, 206, 208, 225, 237, 243, 304, 307–309, 312, 337, 358, and letters in the chapter titled "Marital Relations," 264–292.

56 Ibid., 8.

57 Ibid., 55.

58 Ibid., 304.

59 Ibid., 107, 85, 235, 297. Many women commented on having good husbands who were kind and caring. See 16, 17, 20, 54, 69–70, 106, 235, 383, 408.

60 Ibid., 169, 177. See also 18, 30, 52, 54, 92, 126, 129, 132–133, 206, 208, 300, 399, 402.

61 Florence 1930, 119–120; see also 102–104.

62 Sanger 1928, 69, 104, 119. See also 22, 34, 70, 105, 300–301, 305; Florence 1930, 20.

63 Florence 1930, 104, 103. Letters to Sanger also speak of the dread of sex. See 163, 170.

64 A small but significant number of the letters speak of women's use of abortifacients to terminate pregnancies. These letters express a range of judgments on the practice. Nearly all express a desire to avoid the practice and its deleterious effect on their health. See Sanger 1928, 27, 114, 171, 182–183, 255, 271, 300–301, and the chapter titled "Desperate Measures," 394–410.

65 Ibid., 383, 297–298, 302, 314. See also other letters in the chapter titled "Methods That Fail," 293–321. Although sales of commercial contraceptives in this period grew tremendously, high sales figures should not be taken as a measure of satisfaction with the available methods. Florence 1930, 107, notes that no one *liked* any of the current methods, but "they offer a release from a much more serious consequence."

66 Ibid., 147, 271. See also 145, 302, 356, 378.

67 Florence 1930, 72, 112. On this basis she recommended the condom to married couples instead of the diaphragm.

68 Berlant 2008, 5, 15–16, 19–21, 285; Avery Gordon 2008, 142–143.

69 Sanger 1928, 363–364.

70 Berlant 2008, 3.

71 Sanger 1928, 426.

72 Ibid., 8–9. Many of the letter writers were directly responding to calls for birth control in Sanger's writing, specifically mentioning that they had read her books calling for contraceptive legalization as women's right.

73 Berlant 2008, 27.

74 Howard 1999, 72–73, speaks to the connection between the rise of scientism and the dismissal of sentimentalism.

75 See Kopp 1933.

76 Pearl 1932, 366–367; see also 403, 404, which specifically criticized the statistics of birth control clinic research as simplistic.

77 Ibid., 366.

78 Notestein 1982, 655, claims that Edgar Sydenstricker funded Pearl's research, hoping it would persuade him that fertility was socially determined. See also Reed 1978, 413, on this point. See also Ramsden 2002, 885–887, on the folklore surrounding Pearl's shift in perspective. On the impact of Pearl's research see Ramsden 2002; Ryder 1984.

79 See Haraway 2000, 106; Subramaniam 2014.

80 See Whelpton 1938, 88–93, which identifies investigation into the factors

that shaped desired family size, especially economic, social, and psychological factors motivating voluntary contraception and their social control, as the crucial next steps in the scientific investigation of fertility. Preparation of this book was supported by the MMF, and a presidential committee of the PAA that included Alfred Lotka, Frederick Osborn, Frank Lorimer, and Frank Notestein reviewed and approved it. Whelpton, v, details the participation of Pearl, Notestein, and Osborn in the section on fertility.

81 Pearl 1932, 399, 367 (emphasis in original).

82 Ibid., 367.

83 Ibid., 406, 399.

84 Ibid., 399.

85 Ibid., 400. Pearl ducks the question of nonmarital fertility by simply limiting the sample to married women, because, he argues, the data on illegitimate pregnancies were unreliable.

86 Pearl 1933, 609. See also Pearl 1934a, 356. The only reference to the process producing pregnancy, Pearl's single-sentence description of fertilization uses the sanitized scientific language of the social hygiene movement, which de-sexed sex education in the 1910s. See Simmons 2009, 18, 31–40.

87 Pearl 1934b, 249; 1933, 610. The constant ensures that the equation always yields a positive number. See also Pearl 1934a, 357–358, 366; 1932, 401; 1934b, 254.

88 Pearl 1933, 609. Pearl 1933, 607–608, notes that it is possible to measure risk based on the "rate of sexual intercourse per unit of time." But he rejects this, perhaps because locating pregnancy risk in ovulation effectively distances population science from the topic of sexuality.

89 Pearl 1934a, 400–401. This conceptualization of pregnancy risk enables the aggregation of each woman's entire marital life, regardless of the historical moment in which that life occurred. Thus Pearl's pregnancy risk model is also independent of historical events.

90 Ibid., 356; Pearl 1933, 609. On the standard menstrual cycle as a natural, nonculturally influenced 28-day process, see Pearl 1933, 611. See also Clarke 1998, 4, 70, 123, 158, 191, on the history of the science on menstruation. See Oudshoorn 1994, 136–37, on the use of ovulation in current contraceptive effectiveness studies. See also Takeshita 2011, 17–20.

91 Pearl 1932, 399; 1934b, 262, 252; 1934a, 385, 363.

92 See also Petchesky 1985, 9–10, 35–40, and McCann 1994, 125–127, on this point.

93 Sanger 1928, xiii.

94 Ryder 1984, 8. Stix and Notestein 1940. Although *Controlled Fertility* was coauthored, I attribute the methodology, metrics, and conclusions to Notestein. He is the statistician, and its arguments are completely consistent with his other published work. See also Ryder 1984, Caldwell and Caldwell 1986, and Szreter 1996 on the influence of this study on Notestein's later work and demographic knowledge of fertility in general.

95 Stix and Notestein 1940, 5, 11, 18–19, 22, 159–161, 93.
96 Ibid., 149, 6–7, 63. In such comparisons, the clinic's recommended method proved to be much more effective. But, Notestein, 6, argued, if the use period were less than a year, this method would "classify as 'successes' patients who [had] had no opportunity to fail." By his calculation, 49, an average of four months of non-use was required to produce conception.
97 Ibid., 8, 31.
98 Ibid., 31, 32, 30, 33, 58, see also 44, 168–182.
99 Ibid., 23–25, 42–43.
100 Ibid., 148. See also Ryder 1984, 6–8.
101 Stix and Notestein 1940, 31.
102 Malthus 1989 [1798], 10.
103 Stix and Notestein 1940, 59–60, noted that the effectiveness rates calculated by this method were much higher than those reported by studies that used case-failure calculations. They specifically cite both Kopp 1933 and Hannah Stone 1933. Notestein's calculation of nonclinical contraceptive effectiveness was also almost double that which Pearl had calculated. Pearl's method incorporated a decline in fertility with increased marriage duration, which was missing from Notestein's.
104 Stix and Notestein 1940, 45, 59–60. When douching was used in conjunction with a second method, it was included in the category of the second method (51). Perhaps this helps to explain the high effectiveness rating for the condom and withdrawal, two methods commonly used in conjunction with douches.
105 Ibid., 91–95. The effectiveness rate of diaphragms is hard to find; it appears in a table near the end of the analysis (110). The post-clinic rate for condoms was 95%. The exclusive use rate includes those patients who could not be located, and who were therefore assumed not to have continued to use the clinic method (94–95).
106 Notestein and Kiser 1935, 41; Stix and Notestein 1940, 149.
107 Stix and Notestein 1940, 103.
108 Ibid., 33, 34; Notestein 1982, 654. See also Oudshoorn 1994; Takeshita 2011.
109 *Marital adjustment* was the standard nomenclature used at the time to label satisfactory marital sexual relations. *Satisfaction* is undefined. See Stix and Notestein 1940, 33–35, 57, 89–93, 97. Abortion was not often used as a primary method but as the "last resort" (85); 41% of the sample who experienced an "accidental" pregnancy terminated it through abortion (86).
110 Ibid., 53, table 18. Szreter 1996, 593, notes that this omission would become standard in demographic theory, which presumes that birth control simply replaced "prudential marriage."
111 Stix and Notestein 1940, 101. All the information on partners' reactions came from the women.
112 Ibid., 102. The chart also indicated that a large group found no method satisfactory.

113 Ibid., 153, 2.
114 Ibid., 2.
115 Pearl 1932, 395. His class categories relied on husband's occupation on a scale of unskilled manual labor to skilled professions combined with a rough estimate of family income. His racial categories, "Negro and white," were assigned on the basis of visual perception. Szreter 1996, 350–360, and Susan Cott Watkins 1993, 562, note that this categorization of class obscures gender and class intersections and inscribes the doctrine of separate spheres.
116 Pearl 1934b, 267, 266; 1934a, 391. Pearl 1934a, 400, also concluded that his research made it "plain" that the eugenic program of encouraging more births among the well off was "silly."
117 Pearl 1934a, 370. See also 1934b, 255–256, 268–269, where he noted that during the interviews, "hundreds and hundreds of women" who did not practice contraception "plead[ed]" for "information and instruction."
118 Pearl 1934a, 363, 397–398. For similar wording, see also 1934b, 252, 260–261, 268.
119 Pearl 1934b, 256, 259, 261; see also 1934a, 368–369, 374, 379.
120 Pearl 1934a, 379; see also 366, 381, 385, 398; 1934b, 260, 262.
121 Pearl 1934a, 374; 1934b, 260.
122 Pearl, 1934a, 381.
123 Ibid., 385, 381; 1934b, 261–262.
124 Stix and Notestein 1940, 91, 97, 50, 154.
125 Ibid., 115–117, tables 52 and 53.
126 This research was instrumental in the acceptance of contraception among social eugenicists, for whom universal contraceptive use became a goal by the mid-1930s. Osborn 1940, 149–156, relies on Pearl's research.
127 On cultural competencies and configurations of race see Stoler 1995, 8, 12, 45–53, 90, 205–206.
128 Pearl 1934a, 401.
129 Stix and Notestein 1940, 151. The immediate focus was on the impact of differential fertility between the urban professional class and the rural poor in the United States. On America's rural poor, see 151–153.
130 Ibid., 154.
131 Kopp 1933; Hannah Stone 1933. See also Sanger and Stone 1931.
132 Stix and Notestein 1940, xiii; Notestein 1939b, 125. The first page of the preface states that the project was initiated at Sanger's invitation because she recognized "the need for appraisal of clinic's work by persons unconnected with [the clinic]." The MMF accepted on "the mutually agreeable stipulation that full responsibility for all the details of the investigation and the report should rest with the present authors" (xiii).
133 Stix and Notestein 1940, 5.
134 Ibid., 23, 25, 106–107. The list, 109–110, includes three reasons for the post-clinic improvement in effectiveness: greater determination, better

instruction on physiology and the technique of douching, and providing poor users with better-suited methods.

135 Ibid., 149, 26. As evidence of this higher than normal motivation, Notestein points out that women often came to the clinic after a period of high fertility. That is, for the sample, there was a clear spike in the pregnancy rates in the two years before clinic attendance (88). The sample also displayed greater induced abortion rates during the immediate pre-clinic period, which were taken as evidence of atypical motivation to limit family size (79, 85). These rates were not taken as evidence of either the economic pressures of the 1929–1930 period or the tensions of married life.

136 Notestein 1939b, 124; Stix and Notestein 1940, 107–112, 151, 149.

137 Stix and Notestein 1940, 150.

138 Ibid., 149. Like Notestein and Pearl, Himes began his work in 1930; he distanced himself from propaganda, whose influence he also discounted (Himes 1936, xiii, xiv, xx). Feminist historiography also follows Himes's presumption. Linda Gordon 1976, xii-xiii, 3–7, assumes that the desire to control fertility is enacted in practice, whether or not that practice is successful.

139 Himes 1936, 333–394, cites both Pearl's and Notestein's articles throughout his work. Himes served as an editor of A. P. Pillay's *Marriage Hygiene,* an international sexology journal published in Bombay in this period. See Ahluwalia 2008, 11, 63–65.

140 Himes 1936, xii (emphasis in original). As a social practice repeated in time and space, fertility control was not merely "ephemeral," but was a phenomenon "worthy of scientific inquiry."

141 Ibid.

142 Stix and Notestein 1940, 149. They calculated an unusually high effectiveness rate for withdrawal, 78%.

143 See Himes 1936, 392. Himes also mentions feminism as a factor. Notestein does not.

144 Diffusion, borrowed from the physical sciences, reflects the commitment to scientism within the social sciences of this period. See Ramsden 2002, 868–869; Camic 2007, 276–278; Himes 1936, 209–210, 379–381; Stix and Notestein 1940, 153; Davis 1937.

145 Stix and Notestein 1940, 144–145, 1.

146 Notestein 1939b, 122; 1939a, 511.

147 Himes 1936, 391–392. In this way, Himes gives more credit to the birth control movement than does Notestein.

148 Stix and Notestein 1940, 151.

149 Notestein 1982, 659. The "rubbish" Notestein references reflects both the claim that biology accounted for fertility differences and the claim that access to effective contraceptive methods alone mattered. Sweeping the rubbish away, he notes, "underpin[ned] 'transition theory.'" See also Szreter 1996, 17.

150 Notestein 1939b, 124; Stix and Notestein 1940, 151, 3, 153. See also Stix and Notestein 1940, 144–145, citing Carr-Saunders on the small-family system.

151 Stix and Notestein 1940, 157, 154; see also 155 and Notestein 1939a, 501–502.

152 Clarke 1998, 165, likewise notes that "configuring women as the primary users of contraceptive technologies was, in fact, a core goal of population control groups."

153 Bashford 2014, 329–332, 344–347, notes language of freedom and reproductive choice in what she terms liberal eugenics. However, my reading gives greater weight to the implication in Notestein and his compatriots' articulation of this freedom to act as limited to actions that are congruent with social needs and goals as they defined them.

154 See Takeshita 2011 on biopolitical scripts that organize contraceptives into social worlds and women's lives.

155 See Casper and Moore 2009 for an incisive feminist analysis of infant mortality statistics.

156 Most demographers credit Louis Henry with the theoretical innovation of the natural fertility rate based on research conducted in the 1950s on the white fundamentalist Protestant community in the western United States, the Hutterites. Their high expressed fertility rate was taken as proof of Notestein's estimate. It is important to note that the high fertility of the Hutterites was produced within a social context that valued high fertility, opposed contraception, and supported intensely patriarchal marriage. See Szreter 1996, 26–27, 367–371; Wilson, Oppen, and Pardoe 1988, 3–6, for this history and critique of the concept.

157 Alan Guttmacher, quoted in Takeshita 2011, 33. See also Riley and McCarthy 2003, 74–75.

158 Clarke 1998, 205. See also Smith 1990, 3–4, 68–70.

159 For US demographers in the 1930s, the backward populations were found in the rural United States. See Stix and Notestein 1940, 150–152.

Chapter 5. Demographic Transitions and Modern Masculinities

Epigraph: Appadurai 1996, 120.

1 Thompson 1929, 16; Carr-Saunders 1936, v. Nationalist pressures tore the IUSSPP apart in the late 1930s. See also Bashford 2014, 81–106; Ramsden 2002, 868–869; 2003, 556–557; Connelly 2008, 67, 74–75, 78–79.

2 Carr-Saunders 1936, 1. Notestein 1945b, 37, observed that for Asia, Africa, and Central and South America, estimates "are little better than informed guesses." See also Notestein 1945b, 50; 1944, 432; Kingsley Davis 1956, 310; Symonds and Carder 1973, 7, 78; Riley and McCarthy 2003, 52–53.

3 Taeuber 1946, 254. See also Riley and McCarthy 2003, 75.

4 Kingsley Davis 1945, 1; Notestein 1948, 249–250.

5 On Rockefeller, see Critchlow 1999, 19–29, 161–172; Connelly 2008, 155–157,

162, 165, 169, 329–330. On Draper, see Critchlow 1999, 66–74; Connelly 2008, 186–188, 198, 231–232, 299–300. On Moore see Critchlow 1999, 16–18, 20–33, 66–68, 150–154; Connelly 2008, 186–188, 193. See also Ferguson interview, 1974, 20, 28, 30, 32–35, 40, 46–48, 56–57; Piotrow interview, 2002, 17, 19, 20–30, 40–41, 59, 93–94, 107.

6 Notestein 1968, 554.

7 See Kiser 1981, 494; Van de Tak 1991, 94; Notestein 1969a, 367; Ryder 1984, 5, 9–11; Ramsden 2004, 301, 306.

8 See Ramsden 2004, 283, 290–291, 306; Connelly 2008, 159–160, 170–171; Hodges 2004, 1161; Van de Tak 1991, 5, 94, 333.

9 Porter 1995, 213.

10 See for example Demeny 1988, 466; Hodgson 1988, 554–555; Heer 2004, 60–61; Kirk 1996, 385.

11 See "Methods and Means of Influencing Leadership," August 1957, 1; Dudley Kirk, "Comments of Significance of Sixth Meeting of Ad Hoc Committee," 2, November 26, 1956, RAC-PC, IV3B42, B1, F7.

12 Both men published extensively, often making very similar arguments in multiple publications. Therefore, I cite instances of similar language and logic throughout.

13 Notestein 1982, 660, 664.

14 Ramsden 2004, 283, notes the OPR's role as intellectual gatekeeper.

15 See Van de Tak 1991, 332–333, 335. Clyde Kiser also worked at the MMF throughout this period and facilitated research relationships between it and the OPR. After he completed his degree, Dudley Kirk went to the US State Department, where he worked from 1947 through 1954, the first to hold the position of demographer. Thereafter he served as the director of the demographic division of the Population Council until 1967.

16 He subsequently moved to Berkeley, where he built a stronghold of demographic research in the 1950s and 1960s. See Heer 2004, 73–78, for a list of Davis's students.

17 Van de Tak 1991, 6–7, 43, 93, 232, 328, 332–333.

18 Kiser 1953, 109, notes the importance of the office in shaping demographic knowledge building after 1940. On the Indianapolis study, see "A Study of the Social and Psychological Factors Affecting the Fertility of Protestant Couples," January 2, 1941, RAC-PC, IV3B42, B18, F294. Kiser 1971, 24, notes that Osborn was a driving force behind the study. See Taeuber 1946, 263, on its use of opinion research tools to investigate attitudes and motivations related to family size. It surveyed white Protestants. See also Ramsden 2003, 548, 573–577.

19 Taeuber 1946, 265. See also Van de Tak 1991, 115; Kiser 1981, 491; Ramsden 2004, 291; 2003, 561; Notestein 1971, 74; Bashford 2008, 331–332.

20 See also Notestein 1981, 487, on attention given by the UN Population Commission to the Davis and Taeuber books.

21 Cf. Carr-Saunders 1936, 275. Research for the book was completed while

Davis was at the OPR. He relied entirely on published materials available in the library, only visiting India after completing the analysis, a fact of which he was proud. See Van de Tak 1991, 116.

22 Taeuber's volume built upon work she began in conjunction with a 1948 Rockefeller Foundation–sponsored delegation to the Far East. See Balfour et al. 1950.

23 Symonds and Carder 1973, 43.

24 Notestein 1950a; Van de Tak 1991, 231; Symonds and Carder 1973, 46–47, 52–53; Connelly 2008, 122, 126, 142–143. See also United Nations 1951, 1952, which provide its first population projections and discussion for improving age reporting, respectively. Piotrow 1973, 12–13, reports that "about one-half" of the *New York Times* coverage of population in the 1950s was "directly or indirectly related to UN statistical reports and meetings."

25 Thompson 1929, 320, 331–332, 124. See also Connelly 2008, 139; Hodgson 1983, 7–8. Bashford 2014, 7–9, 50–51, 145, also notes Thompson's anticolonialism.

26 Van de Tak 1991, 118; Kirk 1996, 363. See also Stix and Notestein 1940, 150–152; Notestein 1939a, 499; Notestein 1944; Kingsley Davis 1944. Cf. Szreter 1993, 661; Caldwell and Caldwell 1986, 12; Van de Tak 1991, 117; Connelly 2008, 122–123.

27 See Symonds and Carder 1973, 52, who note that Notestein, Davis, Thompson, Lorimer, Philip Hauser, and Ronald Freeman participated in the meetings.

28 See the references for a full list. Kingsley Davis 1945 was the lead article in an issue devoted to population change that included articles by many prominent demographers. Davis contends that Notestein was miffed that he wrote the lead article. See Van de Tak 1991, 117–118.

29 Carr-Saunders 1936, 96–105, 132, 330. His concept of optimum population allows for the "maximum income level per head," a precursor to the contemporary variable of per capita income. Notestein 1945b, 37, 54–55, uses Carr-Saunders's figures for population change before 1900. See also Bashford 2014, 62–63, 80, 84, 167–168, 307.

30 See Thompson 1929, 107, 132, 145–150. See also Carr-Saunders 1936, 137–144; Bashford 2014, 55–80.

31 Opening the "empty spaces" controlled by Europeans (western United States, Australia, Central and South America, and parts of Africa) for "Chinese, Indian, and Indonesian settlement" was suggested in Thompson 1929 and Mukerjee 1946. For more on Mukerjee see chapter 6. See also Bashford 2014, 107–153, which provides a full discussion of debates about immigration and anticolonialism.

32 Carr-Saunders 1936, 42–43, 59; Notestein 1943, 165; 1945b, 38–39.

33 Kingsley Davis 1951b, 61, 3. See also Carr-Saunders 1936, 62–63, 75, 135.

34 See Kirk 1944, 28ff; Kingsley Davis 1951b, 72–73; 1956, 302; Notestein 1952, 16. See also chapter 4.

35 Carr-Saunders 1936, 320; Kingsley Davis 1951b, 3. See also Notestein 1943, 169, 1945b, 36.

36 This replicated the New Deal–era agricultural policies and would lead to the Green Revolution in the 1960s. See Escobar 1995, 3–5, 21–39; Bashford 2014.

37 Notestein 1945a, 1945b, 1950b, 1952. See also Bashford 2014, 270–271.

38 See O'Connor 2001, 99, 102–106, 107–113, on the Cold War consensus in sociology regarding the family, in which she, 109, notes that patriarchy "became more a necessity than cultural norm."

39 Kingsley Davis 1948, 555–562; 1959. There is also clear tendency of social sciences to treat cultures as types.

40 Balibar and Wallerstein 1991, 55. See also Stoler 1995, 91; Goldberg 1993, 77–81; Cooper and Stoler 1997, 10–11.

41 O'Connor 2001, 107–117. See also Stern 2005, chapter 5; Ladd-Taylor 2001; Balibar and Wallerstein 1991, 49, 56.

42 Stoler 1995, 11, 133–134.

43 Heer 2004, 48, citing Kingsley Davis 1955. Davis defined it as functionless because it exceeded the need of the economy for labor. See also Kingsley Davis 1950a, 102; 1950b, 48; 1951b, 87.

44 Notestein 1948, 251. See also 1951, 99, 128. On gender and functionalism see Stacey and Thorne 1986, 306–307.

45 Heer 2004, 11–115. Kingsley Davis, Parson's student, also participated in the formulation of this theory. Banton 2002, 189, calls Kingsley Davis 1948 "the best textbook of the generation." Part 5 of the book deals with modern population trends. Davis was working on his India and Pakistan book at the same time. He explicitly contrasts the United States and India throughout the book and includes a detailed discussion of the Indian caste system. See Van de Tak 1991, 116–117. On structural functionalism in demography see also Riedmann 1993, 104; Szreter 1993, 673, 684.

46 O'Connor 2001, 106. See also Kingsley Davis 1948, 318, 328–335.

47 Kirk 1944, 28.

48 Stern 2005, chapter 5. See also May 1999.

49 Notestein 1948, 249; Kingsley Davis 1948, 554–555.

50 Notestein 1944, 438.

51 Notestein 1948, 249–250. See also 1944, 431–432; 1950b, 96.

52 Kirk 1955, 18. See also Kingsley Davis 1944, 263; 1951b, 222; Notestein 1945b, 39.

53 Notestein 1948, 250. See also 1944, 431–432; 1950b, 96; 1952, 16.

54 Notestein 1945b, 39. See also 1944, 431; 1945a, 95; 1951, 128; Kingsley Davis 1951b, 64.

55 Notestein 1948, 250–252; 1944, 431–432; 1945b, 39–40; 1952, 16–17. See also Kingsley Davis 1948, 554–555.

56 For insider accounts of transition theory, see Greenhalgh 1996, 36–41; Szreter 1993, 660–663, 666–675; Riley and McCarthy 2003, 42–51.

57 Notestein 1945b, 37, 39.
58 Notestein 1945b, 39, 41; 1948, 250; 1952, 15.
59 Kingsley Davis 1956, 311–312; Kirk 1955, 18; Notestein 1950b, 90–91; United Nations 1951, 5. See also Connelly 2008, 116, for a contemporary version of the story of DDT.
60 Notestein 1948, 250. See also 1950b, 92; 1945b, 41; 1952, 20.
61 Notestein 1950b, 92.
62 Notestein 1948, 250–251. See also 1945b, 39–40; 1951, 128.
63 Notestein 1948, 250; 1944, 432. See also Kingsley Davis 1944, 264.
64 Kingsley Davis 1951b, 79.
65 Notestein 1950b, 96, 97–99; 1952, 16–17; 1943, 166; 1945b, 40–41. For discussion of interventions designed to reduce fertility, see also 1944, 438–441; 1950b, 96–97; 1945b, 41, 52; on Davis's views, see Heer 2004, 115.
66 Kingsley Davis 1944, 264.
67 Ibid., 263–264 (citing Notestein 1943, 165); 1951b, 25. See also Notestein 1951, 128. Cf. Bashford 2014, 309, who likens the metaphor of waste to ecological thought.
68 Notestein 1948, 251.
69 Notestein 1948, 251.
70 Notestein 1950b, 94. Notestein relies on Davis's research in his writings. On Indian mortality decline and growth potential, see Kingsley Davis 1943, 77–78; 1944, 260–263; 1951b, 33–38.
71 Kingsley Davis 1944, 259; 1951b, 28; 1945, 7; see also 1951a, 12.
72 Kingsley Davis 1951b, 28; 1944, 260.
73 Kingsley Davis 1951b, 5, 12, 93, 97, 218, refers to India as peculiar or having a peculiar culture. On differential and/or partial diffusion see 1944, 265–266, 277; 1945, 6, 7; 1951b, 148.
74 Kingsley Davis 1944, 265–266. On "borrowing," see Kingsley Davis 1945, 6; see also Notestein 1950b, 95, who relies on Davis's analysis.
75 Kingsley Davis 1944, 266–269, and Notestein 1944, 432, describe these improvements. The benefits of colonialism constitute a standard feature of British colonial discourse on India. See Nair 2011, 227.
76 Kingsley Davis 1944, 266. See also Notestein 1944, 432; chapter 6.
77 Kingsley Davis 1944, 263; 1951b, 148.
78 Notestein 1944, 433, 431. See also 1948, 250–251; 1950b, 96; 1951, 128.
79 Kingsley Davis 1951b, 224; 1950a, 106. See also Escobar 1995, 41ff, on modernization as an interventionist discourse.
80 Notestein 1951, 101.
81 Immigration, which had been central to the European transition, was curtailed in the West by ethnicity-based legal restrictions. Supports of continued restrictions specifically ruled out immigration as a solution to overpopulation because it would diminish the pressure necessary to spur fertility decline. For instance, Kingsley Davis 1950b, 45–46, rejects migration.

82 Kingsley Davis 1944, 266; 1945, 6; 1946, 253; 1951a, 9; 1951b, 58, 213, 226. See also Notestein 1950b, 94–95; Balfour et al. 1950, 11. On the language of retardation see chapter 3.
83 "Summary," April 4, 1957, 1–2, RAC-PC, IV3B42, B2, F12. See also Mauldin, "Should We Attempt," 2–3, RAC-PC, IV3B42, B2, F13; Notestein 1952, 23.
84 See Kingsley Davis, 1951b, 85–86, on the dependency ratio; see also 1950b, 48; 1950a, 102; 1951b, 219. For a feminist critique of dependency theory, see Bergeron 2006.
85 Notestein 1944, 426.
86 Kingsley Davis 1950b, 45.
87 Kingsley Davis 1951b, 82, 220. See also 1943, 79; 1945, 7; 1946, 254; 1950b, 45.
88 Notestein 1944, 425, asserted that "fertility is close to the biological maximum" in "colonial and semi-colonial areas."
89 Whelpton and Kiser 1945, 112. This is the same edition of the *Annals of the Academy of Political and Social Sciences* in which Davis's introductory essay outlined the demographic transition and first used the explosion metaphor.
90 Notestein 1945b, 41.
91 Notestein 1944, 439.
92 Kingsley Davis 1945, 7. See also Notestein 1944.
93 Kingsley Davis, 1951b, 231. See also Notestein 1951, 100.
94 Balfour et al. 1950, 118.
95 Kirk 1955, 24; Notestein 1944, 439; Szreter 1993, 671. See also Notestein 1950b, 99.
96 Escobar 1995, 39.
97 Notestein 1950b, 96.
98 Notestein 1945b, 40–41. See also Kingsley Davis's views on this point in Heer 2004, 115.
99 Notestein 1944, 432; 1950b, 98. See also 1950b, 96; 1952, 18.
100 Balfour et al. 1950, 117.
101 Notestein 1950b, 97.
102 For additional examples, see Kingsley Davis 1963, 352; 1951b, 226; Notestein 1951, 99–100; 1952, 16.
103 Thompson 1929, 301, uses this phrase in reference to colonial nations.
104 Quite literally, if women appear at all, it is after the individual and *his* motivations are discussed. See for instance Balfour et al. 1950, 80–82; Notestein 1951, 99, 128; 1952, 16; Kingsley Davis and Blake 1956, 223–225.
105 Petchesky 1985, 10; McCann 1994, 126; chapter 4.
106 Kingsley Davis 1951b, 227.
107 Notestein 1950b, 99; 1951, 128; 1952, 27–29.
108 Kingsley Davis and Blake 1956, 218–219. See also Bhatia and Tambe 2014 on demographers' influence on efforts to raise the age of marriage in India in the 1970s.

109 Heer 2004, 48; Balfour et al. 1950, 80.

110 Notestein 1950b, 96, 99; 1944, 440; 1969b, 29. On the position of women in the demographic family, see also Susan Cott Watkins 1993, 561; Stacey and Thorne 1986, 307, 309.

111 Susan Cott Watkins 1993, 556–557, notes that demographers took up the language of women's empowerment expecting women to use the empowerment only to control population growth.

112 Kirk 1955, 24. See also Kingsley Davis 1950a, 101.

113 Kirk 1944, 34.

114 See Kingsley Davis 1951b, 4; see also 152, 216; 1944, 256, 265, 278; 1946, 243.

115 Kingsley Davis 1944, 278.

116 Kingsley Davis 1946, 254; 1951b, 82.

117 Notestein 1948, 253.

118 See Notestein 1939a, 511; 1944, 437; 1948, 254; 1945a, 96; Kingsley Davis 1967.

119 Balfour et al. 1950, 11, 118. See also Notestein 1969b, 29; Connelly 2008, 137. Balfour later worked at the Population Council.

120 Notestein 1944, 437.

121 Balfour et al. 1950, 119–120. See also Notestein 1948, 254; 1950b, 99.

122 An exception to the silence on gendered differences is Kingsley Davis and Blake 1956, 225, who, writing while married to each other discuss the potential for conflict between spouses. This article became the template for much research in the 1960s, as well as the World Fertility Survey in the 1970s, and it remains a classic. See Van de Tak 1991, 120–121.

123 Heer 2004, 48.

124 See Kingsley Davis 1967.

125 Cf. Szreter 1993, 667–679.

126 Notestein 1968, 558. See also Ramsden 2004, 298–299.

127 Van de Tak 1991, 6, and Notestein 1969b, 35, describing the late 1950s and early 1960s.

128 Notestein made an extended trip to India in 1955. See Population Council 1978, 28; Notestein to Osborn, November 21, November 30, and December 27, 1955, RAC-PC, IV3B42, B23, F356.

129 See Notestein 1969b, 27–28, on Population Council and IUD. See also Takeshita 2011, 61, 65, 67.

130 See "Conference on Population Problems," June 20, 1952, RAC, John D. Rockefeller 3rd Papers, RG 5, series 1, B85, F721; Population Council 1978, 9–17; Connelly 2008, 155–160. India figured prominently in these discussions both because of the recent publication of Davis's analysis—he was in attendance—and because it had just become the first nation in the world to endorse fertility control as a matter of national policy.

131 The founding meetings of the council occurred during the same period that IPPF was formally established at an international conference in Bombay that was announced in December 1951 and held in November 1952. The

June 1952 invitation-only conference on population problems held in Williamsburg, Virginia, is generally credited as the founding meeting. However, the council was not incorporated until October of that year and its functions, left open at the June conference, were finalized in November 1952. See Population Council, "Proposed Establishment of Population Council," 100 Years: The Rockefeller Foundation, 3, John D. Rockefeller 3rd Papers, RG 5 series 1.5, box 82, folder 682, http://www.rockefeller100.org/items/show/3750. Accessed August 3, 2015. On the IPPF see Suitters 1973, 47–50, 54–56.

132 See chapter 3.

133 "Proposed Meetings," 3; Irene Taeuber to Frederick Osborn, June 13, 1954, "Philosophy of the Rational Control of Family Size," RAC-PC, IV3B42, B1, F3.

134 See Osborn to Richard Lentz, October 18, 1955, lamenting that some organizations "apparently believed" that "a spirit of controversy is most effective," RAC-PC, IV3B42, B1, F5. See "Proposed Meetings," 1, RAC-PC, IV3B42, B1, F3.

135 Piotrow 1973, 14; Riedmann 1993, 11. See also Piotrow 1973, 15, on Ford grants to the council.

136 Osborn to Blacker, November 21, 1958, RAC-PC, IV3B42, B19, F305; John Durand, acting director, Population Division, to Osborn, November 10, 1953, RAC-PC, IV3B42, B40, F573.

137 "Proposed Meetings," 1, RAC-PC, IV3B42, B1, F3.

138 Osborn 1958, vii, makes it seem as if the Population Council was a participant rather than the organizer that the records clearly indicate it was.

139 See "Methods and Means," 3; Dudley Kirk, "World Population Problems and American Foreign Relations," 2, RAC-PC, IV3B42, B1, F7; and "General Conclusions," May 16, 1957, 1; Osborn, "Summary of Discussion," May 16, 1957, 1, 2, 4, 8, RAC-PC, IV3B42, B2, F13.

140 "Proposed Meetings," January 24, 1955, 3, RAC-PC, IV3B42, B1, F3.

141 Osborn 1958. The Population Council held the copyright and Princeton University Press was the publisher.

142 Committee records show that the papers presented by experts formed the basis of the chapters. The text draws directly from those papers, filtered through Osborn's perspective, which Osborn 1958, ix, acknowledges in the foreword: "In many cases the actual words are those of the competent men and women who took part in the meetings."

143 Osborn 1958, 4–8, 15, 16, 57, 31. He, 60–62, issues the standard call for more research on means and methods, which the council then sponsored.

144 Kirk to Phillips Talbot, February 10, 1956, RAC-PC, IV3B42, B1, F6; Kirk, "Population Problems," 5–6; DK to FO, "Summary of Ad Hoc Meeting, April 11, 1956," April 17, 1956, RAC-PC, IV3B42, B1, F7. See also Osborn 1958, 35.

145 DK to FO, "Summary," April 17, 1956," RAC-PC, IV3B42, B1, F7. See also Osborn 1958, 55.

146 Kirk, "Population Problems," 4, RAC-PC, IV3B42, B1, F7. The rising tide of color refers to Lothrop Stoddard's famous 1920 eugenic treatise, which predicted that differential population growth would cause the collapse of colonialism and white supremacy. Cf. Connelly 2006, 310.

147 William J. Gibbons, "Fertility Control and Catholic Morality," undated, RAC-PC IV3B42, B1, F5; Ashby, "Religious Teaching," RAC-PC, IV3B42, B1, F6.

148 Ferguson was PPFA president from 1953 to 1956 and IPPF vice president from 1959 to 1962. See Ferguson interview, 1974, ii.

149 For the membership, list of discussants, and papers presented see Osborn 1958, appendix, 91–94. Irene Taeuber was the other woman presenter. Leona Baumgartner, physician and professor at Cornell Medical College, who represented the Population Council in India during the 1950s, attended one meeting. See Taeuber, "Cultural and Political Barriers to Collaboration in Policy," March 7, 1956, listed in FO, "Notes on Ad Hoc Meeting, March 7, 1956," 3, RAC-PC, IV3B42, B1, F7; and Ferguson, "The Need for Extension of Effective Practice of Family Planning," May 16, 1957, RAC-PC, IV3B42, B2, F13.

150 Lorimer to Osborn, October 25, 1955, 3, RAC-PC, IV3B42, B1, F5.

151 Minutes, January 11, 1956, 6, 8, 14–15; Philip Ashby, "Religious Teaching and Moral Values," 3, 11–12, January 11, 1956, RAC-PC, IV3B42, B1, F6. See also Osborn 1958, 35, 40, 46–47.

152 See Osborn to Lorimer, Notestein, Warren Nelson, and Robert Snider, December 6, 1956; FO to DK and Robert Snider, December 14, 1956; Kirk to Notestein, Osborn, Parker Maudlin, Warren Nelson, and Robert Snider, January 17, 1957, RAC-PC, IV3B42, B1, F7. See also Osborn to Kirk, February 19, 1957, RAC-PC, IV3B42, B1, F10; Kirk to Phillips Talbot, February 19, 1956, 2, RAC-PC, IV3B42, B1, F6; FO, "Notes," March 7, 1956, 9–10, RAC-PC, IV3B42, B1, F7. See also Osborn 1958, 46.

153 See "Meetings on the Philosophy and Public Understanding of the Personal Regulation of Family Size," April 21, 1955, RAC-PC, IV3B42, B1, F4. Gibbons contributed to the council's anti-advocacy, pro-research perspective. See Minutes, May 11, 1955, 3–4, RAC-PC, IV3B42, B1, F4; Osborn, "Summary," May 16, 1957, 8, RAC-PC, IV3B42, B2, F13.

154 See Osborn to Richard Lentz, October 18, 1955; Lorimer to Osborn, October 25, 1955, 3, RAC-PC, IV3B42, B1, F5. There were exceptions to this point of view. See W. Parker Mauldin, "Should We Attempt to Influence Population Trends," Ad Hoc Committee, May 16, 1957, 5, RAC-PC, IV3B42, B2, F13. Also, both Frances Hand Ferguson and Fairfield Osborn spoke in defense of Planned Parenthood at the March 1956 committee meeting. See FO, "Notes," RAC-PC, IV3B42, B1, F7, 9–10. See also Ferguson interview, 1974, 28–29; chapter 4.

155 Kirk, "Population Problems," 4, RAC-PC, IV3B42, B1, F7.

156 The Eisenhower administration rejected this recommendation. See

Critchlow 1999, 42–45. The council met with Draper. His recommendations reflect the council's imprint in its cautious tone and analysis, which rehearses the demographic transition narrative.

157 See FO, "Notes," March 7, 1956, RAC-PC, IV3B42, B1, F7. On the need for discretion, see Edward Barrett, "Mass Opinion and Population Control," November 16, 1956, RAC-PC, IV3B42, B2, F11; Lorimer to Osborn, October 25, 1955, RAC-PC, IV3B42, B1, F5; Taeuber, "Philosophy," 2, RAC-PC, IV3B42, B1, F3.

158 See FO, "Notes," March 7, 1956, 8; "Methods and Means, Summary," 2, 7–8; Kirk, "Summary," November 26, 1956, 2; "Staff Conclusions," November 16, 1956, 2, RAC-PC, IV3B42, B1, F7.

159 See FO to DK, September 14, 1956, RAC-PC, IV3B42, B1, F10. See also "Methods and Means, Summary," 7–8, RAC-PC, IV3B42, B1, F7; "The United States and World Food Problems: The Malthusian Controversy," undated, 12, 14, RAC-PC, IV3B42, B1, F6.

160 Frederick Osborn, "Summary of VII Ad Hoc Meeting, April 4, 1957," 2, 9, RAC-PC, IV3B42, B2, F12; Balfour et al. 1950, 80.

161 See Osborn 1958, 16, 19, 17. Age of marriage figures track women only. See chapter 4.

162 See "World Food Problems," 10, RAC-PC, IV3B42, B1, F6. See also Osborn 1958, 9–12.

163 Ithiel de Sola Pool, "Methods and Means of Influencing Leadership Overseas: What Social Science Research Has Suggested," undated, RAC-PC, IV3B42, B2, F11.

164 Irene Taeuber to Frederick Osborn, June 13, 1954; "Philosophy," RAC-PC, IV3B42, B1, F3. The initial agenda did not include discussion of the situation in the United States. See DK, "Tentative Agenda," April 12, 1955, RAC-PC, IV3B42, B1, F4; FO to DK, September 14, 1956, RAC-PC IV3B42, B1, F10. See also Osborn 1958, 87–88.

165 India stands out as exemplary of the danger of growth. See Osborn, "Summary," April 4, 1957, 5–7, 2, RAC-PC, IV3B42, B2, F12. "World Food Problems," 11; Kirk to Phillips Talbot, February 10, 1956, RAC-PC, IV3B42, B1, F6. See also Minutes, May 11, 1955, 1–2, RAC-PC, IV3B42, B1, F4; Mauldin, "Should We Attempt," 2, RAC-PC, IV3B42, B2, F13; Osborn 1958, 72, 74–75, 87–88.

166 Osborn, "Summary," May 16, 1957, 5, RAC-PC, IV3B42, B2, F13.

167 Reflecting Osborn's ongoing eugenic commitments, the committee also investigated the effects of differential birth and death rates on the "social and cultural" and "genetic" inheritance. See "Proposed Meetings," January 24, 1955, 3, RAC-PC, IV3B42, B1, F1; Mauldin, "Should We Attempt," 6; Osborn, "Summary," May 16, 1957, 7–8; Osborn, "Family and Social Influences," May 16, 1957, 7, RAC-PC, IV3B42, B2, F13.

168 "General Conclusions," May 16, 1957; Osborn, "Summary of Discussion," May 16, 1957, 8, RAC-PC, IV3B42, B2, F13. Irene Taeuber's editorial assistance was important in moderating Osborn's pejorative tone. See Irene

Taeuber to Frederick Osborn, November 2, 1957, RAC-PC, IV3B42, B2, F12. Cf. Bashford 2014, 330–333, 344–347, 350.

169 Dudley Kirk, "Comments," November 26, 1956, 1, RAC-PC, IV3B42, B1, F7.

170 Ibid. See also Pool, "Methods and Means," RAC-PC, IV3B42, B2, F11.

171 Parry 2013, 4, 49, 77, 88. This theory, Osborn noted, developed before the "scientific machinery" of "its verification." See Osborn, "Family and Social Influences," 2, RAC-PC, IV3B42, B2, F13. Osborn drew on his experience with mass marketing during World War II, where he served as a director of information for the US Army. In that capacity, he advanced survey research techniques through the multivolume *The American Soldier.* See Notestein 1969b; O'Connor 2001, 105–106.

172 See chapter 4.

173 Dudley Kirk, "Comments," November 26, 1956, 2; "Methods and Means, Summary" and "Staff Conclusions," November 16, 1956, 2, RAC-PC, IV3B42, B1, F7. The research on which the committee's approach depended was summarized by two experts at the November 16, 1956, meeting. See Pool, "Methods and Means"; Barrett, "Mass Opinion," RAC-PC, IV3B42, B2, F11.

174 See "Methods and Means, Summary," 7; "Staff Conclusions," November 16, 1956, 2, RAC-PC, IV3B42, B1, F7; Osborn 1958, 58–60, 62–64, 67–68.

175 "Staff Conclusions," November 16, 1956, 2–3; "Methods and Means, Summary," 8, RAC-PC, IV3B42, B1, F7. See also Osborn 1958, 68.

176 On the fellows program, see Notestein 1968, 554, and Population Council 1978, 25–27, 178–193, which lists fellows' names and nationalities. My figures are compiled from that list. See also Riedmann 1993, 106; Osborn 1958, 52–54; Heer 2004, 61.

177 Van de Tak 1991, 95. Kirk identifies "selecting and nurturing Council Fellows" as his greatest contribution to the council. He also discusses the council's support for the work of former fellows in their home nations. Notestein and Irene Taeuber also called for training of indigenous scholars. See Notestein 1950b, 100; 1952, 30; Taeuber, "Philosophy," RAC-PC, IV3B42, B1, F3.

178 These specifically included the CELADE regional center in Santiago, Chile, and the regional center in Chembur, Bombay. See Van de Tak 1991, 97, and Population Council 1978, 194–209, for a list of institutional collaborations by region.

179 Van de Tak 1991, 96. Kirk attributes this sentiment to Bernard Berelson, president of the council in the 1970s.

180 "Excerpt from Remarks by Dr. Irene Taeuber regarding Her Trip to the Orient," January 16, 1954, RAC-PC, IV3B42, B1, F3.

181 The public relations experts consulted were aware of the important role Osborn played in social survey development during World War II. See Pool, "Methods and Means," RAC-PC, IV3B42, B2, F11.

182 Notestein 1968, 559, 554. According to Chesler 1992, 437, the IPPF had far fewer resources in this period, with a budget that did not exceed $35,000.

The council's budget was $1 million per year. See Robbins 1973, 235, on funding between 1968 and 1973. See also Connelly 2008, 206, 232.

183 Piotrow 1973, 13; Notestein 1968, 554. The council supported the IUSSP in this period.

184 Notestein 1968, 555; Population Council 1978, 28–29.

185 Notestein 1968, 558. Kirk credits the council with starting KAP studies in nations of the Global South. See Van de Tak 1991, 96–97. See also Ramsden 2004, 298.

186 Notestein 1968, 558. On birth control public relations campaigns in India, see Chatterjee and Riley 2001.

187 DK to FO, "Summary," April 17, 1956; "Addenda to Report on Ad Hoc Meeting of April 11, 1956," 1–2; Kirk, "Population Problems," 5–6, RAC-PC, IV3B42, B1, F7. See also "Plan for the Sixth Meeting of the Ad Hoc Committee," RAC-PC, IV3B42, B1, F10; R. L. Meier, "Potential Political Consequences of Innovation in the General Area of Fertility," February 1956, RAC-PC, IV3B42, B1, F6; Memorandum, October 26, 1955, RAC-PC, IV3B42, B1, F5; Osborn 1958, 54–56, 65–67.

188 See Berelson 1964; chapters 6 and 7.

189 See Takeshita 2011, 14–15, 17, 38–41, 58, on the basic research and testing for the modern IUD under the council's auspices. See Connelly 2008, 201–206, 216–220, 223–224, 231, 234, 300, on the council's role in IUD development. See also Suitters 1973, 210, 214, 351–352; Ramsden 2004, 297.

190 Notestein 1969b, 28. See also Connelly 2008, 205.

191 Notestein 1969b, 29; 1968, 556, 558. The IUD was recommended in the Taiwanese and Korean clinics sponsored by the council.

192 The Population Council was a public critic of Planned Parenthood's pill. See Piotrow interview, 2002, 46–47.

193 Notestein 1968, 556–557; see also 1969b, 27–28, which notes it is highly effective compared to the pill because "there is no problem of patient neglect." See also Berelson 1964, 7–9. Notestein 1969b, 30, however, also opposed the drastic governmental coercion advocated by Kingsley Davis 1967, saying it was "more likely to bring down governments than birth rate."

194 Alan Guttmacher 1969, as cited in Takeshita 2011, 33. Takeshita provides an incisive analysis of the co-configuration of IUDs and IUD users as biopolitical objects in midcentury population control politics. Similarly, the spokesman for Searle Laboratories identified the pill they produced as "doing mankind a great favor." See Elizabeth Watkins 1998, 112.

195 I borrow this phrase from Escobar 1995, 5, who uses it in reference to the concept of development as a solution to world poverty. The demographic transition helped assemble that linkage.

196 Parry 2013, 47–48; Piotrow 1973, 12–13, 24–25, 48–50; Kingsley Davis 1957, 1959, 1948. See also Notestein 1951, 1959.

197 Notestein 1952, 25.

198 Ahmed 2004a, 118–121, on affects and action.

199 McIntosh and Finkle 1995, 234. See also Riley and McCarthy 2003, 42–51. Szreter 1993, 661, 693, indicates that Notestein 1952, 17–18, published historical data for France and Bulgaria that "directly refuted the theory's model of change." Notestein's protégé, Ansley Coale's, took over at the OPR when he went to the council in 1958. Under Coale's leadership, the OPR carried transition theory into the massive twenty-year European Fertility Project, which sought to identify the timing, causative factors, and overall shape of the demographic transition in over 600 administrative units of Europe but which was unable "to offer any conclusive support" for the theory. The best that could be said is that mortality fell first. Szreter 1996, 25, concludes, however, that the project "grant[ed] to the idea of demographic transition a scientific life after death."

200 The temporal distancing of transition theory, structured as a comparison between their now and our yesterday, distracts attention from the simultaneous baby boom in North America and Oceania. See Coale and Hoover 1958, 17, 13.

201 Kirk 1996, 385. See also Szreter 1993, 686, with whom Kirk disagrees.

202 See, for example, Display Ad for *CBS Reports: Population Explosion*, *New York Times*, November 11, 1959, 71.

203 See Escobar 1995, 33, which notes that Western nations did not deliver on the investments they had pledged to the Global South.

204 Critchlow 1999, 50–53; Connelly 2008, 209–210, 212–213, 220–222, 225, 227–228.

205 "Summary," April 4, 1957, 6, RAC-PC, IV3B42, B2, F12.

Chapter 6. "Second Sight" and "Fictitious Accuracy to the Numbers"

The quotes in the chapter title refer to Gyan Prakash 1999, 34, and Coale and Hoover 1958, 260, respectively. Epigraph: Chand 1939, 40.

1 See "Conference on Population Problems, June 20–22, 1952," 2, RAC-Office of the Messrs. Rockefeller Records, series O, B1, F4.

2 Government acts between 1919 and 1935 devolved governance authority over key aspects of public welfare, including education, health, and sanitation, to provincial Indian legislative bodies. See Prakash 1999, 144ff. See also Arnold 2006 and Hodges 2006a, 2006b, and 2008 with regard to the importance of these acts for birth control advocacy.

3 Zachariah 2001, 3690.

4 Hodges 2006b, 12; Arnold 2006, 23. See also Nair 2006, 68, 74.

5 Chatterjee and Riley 2001, 822–823.

6 Prakash 1999, 49–85, argues persuasively that the Western-educated elite engaged in a process of translating scientific knowledge for their own purposes. Population knowledge was a critical part of this translation project. Reading Indian population discourse merely as colonial imposition thus misconstrues the agency of the Indian elite.

7 The Madras Neo-Malthusian League was founded in 1928 by prominent Tamil Brahmin men, while the anti-Brahmin social movement in the area included birth control as part of its program. See Ahluwalia 2008, 28–30; Hodges 2008, 47–49, 68, 84–93, 99–103.

8 Hodges 2008, 1, 105–138. See also Ahluwalia 2008, 66–70.

9 Caldwell 1998, 688, notes it was founded by the maharajah of Mysore, who was a member of the student eugenics association at Madras Presidency College in the 1910s along with Dhanvanthi Rama Rau, who would found the Indian Family Planning Association and serve as copresident of the IPPF with Sanger.

10 See Ramusack 1989, 37, 42, 45, 51; Hodges 2008, 51.

11 It was banned by US customs as a result. See Ramusack 1989, 37, 45, and Ahluwalia 2008, 6, 11, and 31, on Pillay and his journal. See Hodges 2008, 49–50, on the Madras Neo-Malthusian League's engagement with birth control internationally.

12 More than 1,000 letters from Indian correspondents between 1922 and 1935 are among Margaret Sanger's archived papers. See Chesler 1992, 560, n. 3; Connelly 2008, 90. For an example of articles about India in movement publications, see Mukerji 1933.

13 Indian population scholars were invited to, attended, and made presentations at the Geneva conference in 1927 and related conferences throughout the 1930s. See Sanger 1927, 114–118; P. C. Joshi 1986, 1459; Bashford 2012; 2014, 99, 102, 119–120.

14 On population and governmentality see chapter 2. Cf. Hodges 2008, Ramusack 2006, and Ahluwalia 2008, all of whom argue that there is a disjuncture between prewar birth control advocacy and postindependence population control. Hodges is insistent on the one not presaging the other. Ramusack and Ahluwalia are less so.

15 See also Hodges 2008, 135; Ramusack 2006, 77.

16 Hodges 2008, 141. She credits Kingsley Davis for this representation.

17 Most of the individuals that I identify in relation to Indian population science are characterized in other studies as birth control advocates and eugenicists. I identify them as scientists first and foremost throughout to reinforce their position as scholars and the social science expertise of Indian men involved in birth control advocacy. Cf. Ahluwalia 2008.

18 Hodges 2006a, 115; 2008, 39. See also Arnold 2006, 28; Prakash 1999, 10, 98, 127, 134–137; Ahluwalia 2008, 24–26; Connelly 2008, 29, 32.

19 Zachariah 2001, 3698.

20 Kingsley Davis 1951b, 3.

21 Notestein 1944, 426; 1945b, 50; 1951, 101; Caldwell and Caldwell 1986, 13–15.

22 Bhore 1945, 479–480. See also Arnold 2006, 44–46. Caldwell 1998, 689, notes that Davis and Whelpton were consultants to Nehru on the family planning portion of the first five-year plan.

23 See Symonds and Carder 1973, 37, 52; Reed 1978, 282; Szreter 1996, 24; Connelly 2008, 435, n. 61, on the influence of Coale and Hoover's analysis. See Van de Tak 1991, 115, and 116; Caldwell 1998, 687; and Hodges 2004, 1162, on the influence of Davis's book.
24 Coale and Hoover 1958, 47.
25 That acknowledgment often appears at a moment intended to shame the United States for its hesitancy on policy. It also functions as a shield against charges of ethnocentrism.
26 See, for example, Ledbetter 1984, 737–738, 742, where she cites the *New York Times* on India's effort to reverse the natural order of modernization and population change.
27 Joshi 1986, 1456; Thakur 2012, 96, 104. "The Testimony of Figures" is from Wattal 1916, 11, 7.
28 Chand 1944, 21; see also 1939, 315.
29 See for example Chand 1944, 24.
30 See Appadurai 1996, 126–128, 130; Greene 1999, 33; James 1979; Winch 1987, 13; Caldwell 1998, 682–683.
31 See Hodges 2008, 22–24; Prakash 1999, 130–35; Chand 1939, 121–122.
32 Arnold 2006, 27, 33, quoting John Megaw, director general of the Indian Medical Service.
33 Ibid., 24; Prakash 1999, 128, 260, n. 13, notes that before the mid-nineteenth century, the British looked on the people as part of the landscape.
34 Caldwell 1998, 684. See also Notestein 1944, 426; chapter 5.
35 Social and economic interventions did occur, such as legal changes to the status of girls and women, smallpox vaccination and plague management, and changes in land ownership and forest-use rights. See Ahluwalia 2008, 116–117; Prakash 1999, 138–143.
36 Famine and epidemic traveled along colonial trade routes, causing millions to die in India and China. Connelly 2008, 28–30, citing Mike Davis 2001. See also Prakash 1999, 137–138, on colonial practices that spurred the 1867 cholera epidemic in northern India.
37 Prakash 1999, 128–142; Nair 2011, 237; Hodges 2008, 8–9, 22, 35–37.
38 Caldwell 1998, 684, 687. See also Arnold 2006, 24; Rao 2004, 108.
39 As cited in Chand 1939, 297, who also notes that the alarm was sounded in "every annual report after 1931."
40 The 1931 census report praised the Madras Neo-Malthusian League. See Hodges 2008, 49.
41 Arnold 2006, 27–28, 23; Hodges 2008, 10, 39, 13–14. See also Chand 1939, 6–9, who presents both sides of this argument in sharply written prose.
42 Prakash 1999, 49–52.
43 Ibid., 34.
44 Ibid., 127, 144–147. See also Zachariah 2001, 3698. As an example see Chand 1939, 223–224.
45 Mukerjee 1938, 1.

46 Joshi 1986, 1464. Radhakamal Mukerjee's life's work is an example of the subjectivity and agency Prakash describes. He embraced Western science and sociological theory, founding the Lucknow School of Economics and Sociology. But he also rejected the West's model for India's development, making a long study of Hindu cultural forms. See Joshi 1986, 1458. See also Nair 2006, 67–68.

47 Thakur 2012, 95. Thakur argues that Mukerjee is among the first Indian social scientists to question Western models and that he influenced Gandhi's thinking. But he also notes, 97, that his writing "ends up romanticizing, idealizing, and constructing a useable past at the service of an incipient nation"; see also 96–104.

48 Ibid., 97, 95–99.

49 Arnold 2006, 28; Ramusack 1989, 56.

50 Caldwell 1998, 686. See also Ramusack 1989, 36.

51 Ahluwalia 2008, 29–30; Nair 2006, 70.

52 Wattal 1916, 1–2, 26; see also Chand 1939, 42.

53 Ahluwalia 2008, 33–34; Nair 2006, 70–71, 125, 73.

54 Wattal 1916, 9, 29. Cf. Ahluwalia 2008, 33–34, who reads Wattal as endorsing moral restraint only.

55 Nair 2006, 71.

56 Chand 1939, 47, 48. On Chand, see Nair 2006.

57 Hodges 2008, 13–14; 2004, 1159; Nair 2006, 252.

58 Chand 1939, 214.

59 Chatterjee and Riley 2001, 822. Degeneration was a staple trope of eugenics throughout this period, but care should be taken in reading Indian concern with degeneration as simply reflecting eugenic ideas. Prakash 1999, 89–92, 179–183, 201–203, and 224–226, persuasively argues that the trope of degeneration was central to the nationalist discourse that sought to translate modernity for the Indian nation as the recuperation of a lost past greatness. As an example, see Chand 1939, 216–219, 315, 347.

60 Mukerjee 1931, i–ii; 1938, 47. He takes the "immutable forces of nature" phrase from Raymond Pearl, dismissing it as a "somewhat mystical hypothesis." On the vicious circle metaphor, see ibid., ix, 34; see also Huxley in Chandrasekhar 1961 [1955], 12; 1967, 27.

61 Chand 1939, 363.

62 Chand 1939, 2–3, 183; see also 15–16.

63 Mukerjee 1938, xiii, 217.

64 On Indian perspectives on adverse sex ratios, see Chand 1939, 108–109; Mukerjee 1938, 46–47, 229–232, 235–237; Nair 2006, 69; Chandrasekhar 1961 [1955], 37; 1967, 7. Note that the natureculture of sex ratios was established in this era. That is, the so-called natural ratio of 105 males to 100 females is based on 300 years of parish records from Sweden, understood as recording biologically determined ratios. On arguments about missing women predicated on these sex ratios see Greenhalgh and Li 1995.

65 Wattal 1916, 22–23; Prakash 1999, 88. On the Sarda Act of 1929, which raised the age of consent, see Sinha 2006b, 154–172; Hodges 2006b, 12; 2006a, 128–129; Arnold 2006, 46; Aryee 2006, 230. On widow remarriage see, for example, Mukerjee 1938, 51, 225–228.

66 Hodges 2008, 36.

67 Chand 1939, 94.

68 See Hodges 2008, 26, 35–38, 42, on the 1919 Government of India Act and Madras health centers.

69 Nair 2006, 102; Sinha 2006b, 8–9, 25–26, 43–45.

70 Wattal 1916, 26. At the same time, nature is personified as feminine and merciless, with "nobody to thwart her wishes" (22). See also Arnold 2006, 34; Sinha 2006b, 8–9.

71 Chand 1939, 134.

72 Ibid., 368.

73 Thakur 2012, 99; Joshi 1986, 1460. See also Mukerjee 1941a, 142.

74 See Hodges 2006a, 116–117, 124–125. See also Chand 1939, 367–368.

75 Mukerjee 1941a, 144. See also Mukerjee 1938, 229–237, on the biosocial components of India's sex ratio. Both Ahluwalia 2008, 41, and Bashford 2012, 619, cite his references to "subaltern groups" as "breeding like field rats, rabbits, and fruit flies." Ahluwalia labels it an elitist "Malthusian slur." Bashford argues that this list is indicative of Mukerjee's ecological perspective, in which man is an animal within a complex natural and social environment. I concur with Bashford. Moreover, I find that his language resonates with that of Osborn's social eugenics. The phrase in question appears in his account of the natural checks on population, which he describes using Raymond Pearl's standard animal models—rodents and fruit flies. See Mukerjee 1931, ii, vii; see also 1938, 42.

76 Chand 1939, 11, 134, 10; see also 41.

77 Ibid., 133.

78 Bashford 2014, 240, argues as well that there is far greater continuity between eugenics and Malthusianism than acknowledged by historians.

79 Chand 1939, xi.

80 Mukerjee 1938, xiv, 221, 223. See also chapter 2.

81 Wattal 1916, 16. Mukerjee also held negative views of Muslims. See Mukerjee 1938, 213, where he blames Muslim invasion for the advent of "infant marriage" among Hindus.

82 Chand 1939, 149, 155, 156. But, he says, it is even less warranted in India than in other countries.

83 Ibid., book title, 149. See also Hodges 2008.

84 Prakash 1999, 35, 39; see also 40–46, on "the disturbing presence" of the subaltern "in dominant discourses."

85 As quoted in Hodges 2008, 70.

86 Hodges 2006a, 115–117, 126, 132–134. See also Hodges 2008, 14–16; Nair 2006, 97–105. There is, however, attention given to the differential growth

of the Hindu and Muslim communities. See Wattal 1916, 4–5. After the 1935 Government of India Act allocated electoral representation by religious and caste population proportions, differential growth rates between Hindus and Muslims acquired greater political importance. These communal politics intersected with gender politics to the extent that differential sex ratios affected growth. See Nair 2006, 224–229; Mukerjee 1938, 225–229.

87 See Mukerjee 1938, xiv. See also Wattal 1916, 24–25.

88 Mukerjee 1938, xiv–xv.

89 Ahluwalia 2008, 28.

90 Mukerjee 1938, 216–217, and Chand 1939, 349, praised Sanger and Edith How-Martyn.

91 Nair 2006, 116, 147–148, 190–196, 201–202. See also Ramusack 1989, 36, who notes that Karve lost his position as a mathematics professor as a result. He was charged with obscenity for publishing birth control advice in the vernacular *Samaj Swashtya.*

92 Chand 1939, 149; see also 147–148, 190–194, 201–202, 326–327, 352, 359.

93 Mukerjee 1941b, 792; 1938, 217, 216, xiv. See also Nair 2006, 82–84.

94 Chand 1939, 94; Hodges 2008, 62–63; see also 9–10.

95 Mukerjee 1938, 211–212, 217–218.

96 Chand 1939, 40, 162, 358. See also 325–359, where he talks about birth control as a revolution in human affairs, and 332, 336–337, and 146, where he writes of sexual love in marriage and women's reproductive situation with a feminist inflection.

97 They also shared a negative view of women's competency. See Ramusack 1989, 39, and 2006, 55–57, 64, 78; she notes that the head of the Madras Neo-Malthusian League reported that there were no competent women in India involved in birth control issues.

98 Ramusack 1989, 41–45; Chatterjee and Riley 2001, 821; Hodges 2008, 101–103; Ahluwalia 2008, 90–98, 105–108; Arnold 2006, 39–41. See also Sinha 2006b, 55–58, 60–61, 154–156, 162–164, and 200–201, on women's political activities in this period.

99 Hodges 2008, 102; Ramusack 1989, 48–50; 2006, 75–76; Ahluwalia 2008, 19, 161–162.

100 Arnold 2006, 35–37. However, women flocked to lectures on health and birth control when available to them. See Prakash 1999, 37. Ramusack 2006, 79–80, and Hodges 2006a, 115–117, note that women doctors' use of eugenic language allowed distance from the sexuality that birth control implied. Elsewhere, I make a similar argument about eugenic language in the US birth control movement; see McCann 1994, 122–134.

101 Hodges 2008, 102; see also Ahluwalia 2008, 95.

102 Prakash 1999, 127; Chand 1939, 296–297, 328–329. Chand also notes that Gandhi's views speak to the suspicion that the population question was merely a tool of the British to avoid political and economic change. See also Caldwell 1998, 687.

103 On the exchange between Sanger and Gandhi, see Aryee 2006, 227–234; Hodges 2008, 58; Ahluwalia 2008, 75–77; Ramusack 1989, 50–51. See also Sinha 2006b, 107, on Sanger's explicit anticolonial statements while in India.
104 Chand 1939, 329–341, offers a sustained critique of Gandhi's position. See also Nair 2006, 239–240, who notes that Karve wrote a scathing critique of Gandhi's views.
105 See Nair 2006, 27, 172–175, on the Lucknow conference.
106 Ibid., 167–168. Raja was a member of the Royal Statistical Society and served in the public health establishment in India, funded in part by the Rockefeller Foundation (44).
107 Ibid., 214–215, cites a laudatory article on the exhibition published in the *Times of India,* April 18, 1938. Prakash 1999, 24–25, 34–40, notes the pedagogical power of public exhibits in this period—so much so that the Indian National Congress began to present exhibitions in conjunction with its annual meetings after 1901.
108 Prakash 1999, 197. On the ambiguous impact of these acts on the status of women, see Sinha 2006b, 20–22, 40–41, 50–53, 56–58, 197–199.
109 Nair 2006, 242; Bose 1997, 207. D. G. Karve also served on the Population Sub-Committee along with Sir Vepa Ramesam, the leader of the Madras Neo-Malthusian League. See also Nair 2006, 251–254, on the subcommittee's work.
110 The effect of Britain's wartime redistribution of food on the famine should not be underestimated. See Ó Gráda 2009, 160–184; Amartya Sen 1994. See also Chatterjee and Riley 2001, 822, on the famine inquiry commission's call for population control. See Nair 2006, 254–262, on continuing food shortages in the 1940s.
111 Chand 1944, 28–29; see also 30–31, on contraception.
112 Ibid., 24, 28; 1939, 162–164.
113 Nair 2006, 248–249; Thakur 2012, 90.
114 Connelly 2008, 143. These standards were developed when Notestein and Pascal Whelpton led the UN population division.
115 Chandrasekhar 1959, 28–29.
116 Chand 1944, 10.
117 Chand 1939, 94, 150, 116; Mukerjee 1938, xiv.
118 See Chatterjee and Riley 2001, 826; Ledbetter 1984, 738; Connelly 2008, 146; Rau 1977, 257. See also Ramusack 2006, 55; Hodges 2006b, 16; W. T. Pommerenke to Frederick Osborn, October 18, 1955; DK to FO, June 14, 1955; FO to Frank Notestein, Dudley Kirk, and files, March 29, 1955, RAC-PC, IV3B42, B9, F101; Notestein to Osborn, November 21, 1955, November 30, 1955, and December 27, 1955, RAC-PC, IV3B42, B23, F356.
119 See Nair 2006, 262. Presaging the future trend, Abraham Stone 1953, 212, notes that he "found widespread interest in sterilization throughout India, not merely as a eugenic measure but as a method [of contraception]" and

that the procedure was "resorted to rather frequently" in "some" of the centers he visited in 1951.

120 Connelly 2008, 199. The crowd quotation is from Nehru, who further noted that those Americans "nosed about far too much." Conversely, Connelly 2008, 200, observes that the Population Council staff felt that Indian officials had too much control of their activities.

121 Chand 1939, 94, 325; see also 331–338 for sentimental prose on sexual love.

122 Mukerjee 1938, 219.

123 Chand 1939, 162.

124 Ibid., 94. See Chatterjee and Riley 2001, 813–815, for representations of happy small families on Indian family planning posters.

125 Mukerjee 1941a, 142. Ironically, he asserts his disagreement with Malthus here, arguing for community accountability along with individual accountability for the care of children.

126 Chand 1939, 42; see also Hodges 2008, 50; Chatterjee and Riley 2001, 831. Nair 2006, 102, insightfully observes that the nationalist project was one in which "Mother India . . . would be set free by her sons."

127 The subhead refers to Kingsley Davis 1951b, 220; 1950b, 45.

128 Kingsley Davis 1951b, vii. Chapters appeared in academic journals beginning in 1944, and Notestein 1945b, 50, relies on Davis in his discussions of India's demography.

129 Kingsley Davis 1951b, 4. Note the feminine pronoun.

130 Symonds and Carder 1973, xiv; Coale and Hoover 1958, title page. The 1881–1941 censuses covered the territory controlled by Britain and allied princes, and those boundaries shifted several times.

131 Kingsley Davis 1951b, 16. Davis's analysis exemplifies the slippage between individual and aggregate by which the demographic hierarchy of nations is indexed. See also Hodges 2004, 1157.

132 Davis voices a longstanding lament about the lack of statistics in India. See Alborn 1999, 79–81, 85, which notes, 70, the routine gesture by colonial census administrators of bemoaning the low quality of data and their own heroic mathematical efforts to make "meaningful generalizations in spite of the patently insufficient evidence." See also Chandrasekhar 1959, 42–65, and Rao 2004, 16, who offer similar laments.

133 Kingsley Davis 1951b, 6.

134 Kingsley Davis 1951b, 6; 254–258. His bibliography also includes many of the Indian social scientists discussed above, including Raja, Wattal, Karve, Pillay, and Mukerjee.

135 Ibid., 3.

136 Ibid., 16, 23, 24. He uses the colonial conjecture of the figure of 125 million for that time period. See also Hodges 2008, 141; 2004, 1157; 2006b, 2.

137 Kingsley Davis 1951b, 25, 28; Chandrasekhar 1959, 28–29. After the noncooperation effort of the 1930 census and the war-impacted 1941 census, the

1951 census was the first enumeration in twenty years that Indian demographers expected to be complete.

138 Davis 1943, 77. See also Kingsley Davis 1944, 259–260; 1951b, 28.

139 Kingsley Davis 1951b, 28.

140 The practice of comparing increments of Indian population growth to the growth of small European countries rehearses a trope common in British colonial discourse, where the comparison was regularly made to England and Wales. See Alborn 1999, 69; Kingsley Davis 1951b, 28; see also Ledbetter 1984, 736.

141 With regard to measures of modernity, US demographers after World War II specifically contrasted the demographic situation in India to that of the United States, displacing Britain but maintaining the positional superiority of the West. See Davis 1951b, 28.

142 Kingsley Davis 1951b, 51, 221, 64, 220. He does acknowledge that mortality in India was still very high overall (53).

143 Ibid., 87; see also appendix F, 246, for the method of calculation. The net reproduction rate was 1.09 in 1901 and 1.30 in 1941 (87). See also Nair 2011, 234; Arnold 2006, 28, 30–31; Bashford 2014, 293–294.

144 Kingsley Davis 1951b, 89; see appendix G, 247.

145 Ibid., 213, 205–206, 215–216. On the use of this language, see chapter 3.

146 Ibid., 86, 218. The young age of the population also increased its future growth potential as mortality dropped, because more children would survive and reproduce. For a feminist critique of dependency theory in development economics, see Bergeron 2006.

147 Kingsley Davis 1951b, 215–216, 218.

148 Ibid., 216. Davis's discussion of caste is another example of the continuity between his analysis and the colonial discourse that displaced nationalist perspectives. His book appeared at the same moment that the postcolonial census rejected the categorization by caste that had been central to the British census administration. As the minister for Home Affairs noted at a conference of census superintendents, "Formerly there used to be elaborate caste tables which were required in India partly to satisfy the theory that it was a caste-ridden country and partly to meet the needs of administrative measures dependent on caste divisions" (Chandrasekhar 1959, 26–27, 30).

149 Kingsley Davis 1951b, 221.

150 Ibid., 82, 230–231; see also 1943, 79; 1946, 254.

151 Coale and Hoover 1958, viii–ix, v. The trip took place in the autumn of 1955.

152 Osborn 1958, 9–10. Since it was a non-academic text, Osborn does not provide a specific citation, but the data summary is unmistakable. Coale and Hoover 1958, viii, note that "300 mimeographed copies of a preliminary draft of the book were circulated privately."

153 Coale and Hoover 1958, v, 4. Italics are added to highlight the slippage between individual and aggregate, personal and national, volition.

154 Ibid., 228, 5, 288 (emphasis in original), 286; see also 291, 334–335. Although

they were definite about this figure, they reiterated (4, 259, 278) that none of their economic projections should be taken as forecasts.

155 Ibid., v, 4.

156 See Szreter 1996, 24, 26–27, 368–370, 391, 542, on the high regard for Coale's refinement of Lotka's equations, and Notestein 1982, 666–667, on the use of age-specific rates in population predictions.

157 They note nationalists' aspirations in the first paragraph. See Coale and Hoover 1958, 3, 30–31.

158 Hacking 1992, 140. See also Danby 2012, 504–505, on the "robust confidence in numerability" in mid-1950s economic surveys.

159 See Coale and Hoover 1958, 43–45, appendix A, 337–365, where the procedures are outlined.

160 Kingsley Davis 1951b, 246, follows a similar method of working back from survival rates in reconstructed life tables.

161 Demographically, *infant mortality* refers to the death of infants within the first year of life. The infant mortality rate is the number of deaths per 1,000 live births in a year, excluding miscarriages and stillbirths. See also Chandrasekhar 1959, 72–76, on these figures.

162 Coale and Hoover 1958, 52–53. They used models for 200 and 250, but for charts and graphs they represented the midpoint of 225.

163 Ibid., 50, 54.

164 Ibid., 343.

165 Alborn 1999, 66. See also Coale and Hoover 1958, 31–33. Life tables were originally developed as a tool for accurately pricing life insurance.

166 Coale and Hoover 1958, 351. The other choice, made by census officials, was to assume a higher childhood mortality rate for girls. Their choice thus meant that the life table they constructed did not account for this sex differential in childhood mortality. See Chandrasekhar 1967, 7.

167 Mukerjee 1938, 14, also makes this point. See also Mukerjee 1938, 44–45, 229, 233–235, on the low sex ratio.

168 See chapter 2.

169 Coale and Hoover 1958, 47, 15, 45–46. For the formal equations see 338–344ff.

170 Ibid., 46, 9–13, 56.

171 Ibid., 47, 31.

172 Ibid., 16.

173 Ibid.

174 Ibid., 16, 47, 57–60.

175 Ibid., 59, 30–31, 40, 43–44. Coale and Hoover also discount the earlier figures generated by the Public Health Commission of British India. See also Nair 2011, 244; Chandrasekhar 1961 [1955], 96.

176 Coale and Hoover 1958, 2, 286.

177 Ibid., 56, 60, 61. They rule out both abortion and abstinence as methods, however. They also again point to decline in the West and Japan.

178 Prakash 1999, 12.
179 Coale and Hoover 1958, 288. They assume that a steep decline in mortality is realistic based upon the precedents of past mortality declines through public health measures, again specifying the success of DDT spraying in Sri Lanka in 1946 (62–69).
180 Ibid., 71. Cf. Caldwell 1998, 691.
181 See Ledbetter 1984, 740, 750, as an example.
182 Coale and Hoover 1958, 36–37, 24; see also 22–25. The funds simply go to the expansion of existing facilities rather than to the investment in new and better techniques. They do not explain why additions cannot also be improvements.
183 Ibid., 36, 39 (emphasis in original).
184 Notestein 1952, 23, suggests that a 4% growth rate would not be adequate to absorb the high growth rates of the Global South.
185 Notestein 1952 also relied on the per capita income measure. On its history, see Escobar 1995, 22–24.
186 Coale and Hoover 1958, 260–262. Similarly, ignoring the wealth of non-monetized activity, the World Bank first determined that two-thirds of the world was poor in 1948. See Escobar 1995, 24.
187 See Rao 2004, 126, for a critique of its use for demographic statistics.
188 See Symonds and Carder 1973, 17–18, on the beginning of the shift in the definition of overpopulation from population density to economic development in depression-era debates about immigration. See also Bashford 2007, 2008.
189 Alborn 1999, 63.
190 When the contingency of the figures is acknowledged, it serves only to amplify the sense of crisis because inaccuracies are said to underestimate the "real" problem. See also Hodges 2008, 39–40, and Curtis 2001, 33, on this point.
191 Coale and Hoover 1958, 260.
192 Ibid., 4.
193 Notestein 1945b, 53; 1969b, 9–10, 14; see also 1945b, 37; 1943, 167; 1952, 24.
194 Kingsley Davis 1951b, 90; see also 87–89.
195 Coale and Hoover 1958, 227 (emphasis in original).
196 Szreter 1996, 24, notes that Harvey Leibenstein incorporated their analysis in his theory of economic development. The conclusions reverberate even now in histories of the period. See for instance, Metcalf and Metcalf 2006, 198–199, 256. In demography, doubts about their conclusions only appeared in the 1980s, after fertility declined and economies grew in the Global South. See Hodgson 1988, 558.
197 Jawaharlal Nehru, quoted in Connelly 2008, 199.
198 Chatterjee and Riley 2001, 823. See also Nehru's foreword to Chandrasekhar 1961 [1955].
199 As quoted in Ledbetter 1984, 740–741.

200 Connelly 2008, 197. The resolution was defeated through parliamentary maneuvers coordinated by the Vatican representative, even though they too acknowledged that rapid population growth was a problem.
201 Ledbetter 1984, 742.
202 The mid-1960s was the period of the most intense involvement of US advisors. See Chatterjee and Riley 2001, 824, 826.
203 Chand 1939, 326–327.
204 Chatterjee and Riley 2001, 823. Karan Singh was minister of health and family planning.
205 Ibid., 824. See also Connelly 2008, 318–326. The program involved primarily compulsory sterilization for men and is often cited as one reason why the government fell in 1977. But sterilization had long been a focus of family planning programs in India. See Berelson et al. 1966, 801–804; Abraham Stone 1953.
206 See Chatterjee and Riley 2001, 823, on the "multiple ambivalences" of "India's nationalist population agenda."
207 See, for example, Connelly 2008, 317–326.
208 Prakash 1999, 50.
209 John D. Rockefeller III and Viond Patel, a Gujarati state official, as cited by Connelly 2008, 296. See also Prakash 1999, 71.
210 The fortunes of US demographers would change in the 1980s after the election of Ronald Reagan on a pro-life platform. Connelly 2008, 353–355.

Conclusion

The chapter title refers to Berelson 1969, 541. Epigraph: Chand 1939, 326–327.
1 On the Draper Committee see Critchlow 1999, 41–45; Connelly 2008, 186–187, 198.
2 Smith referenced the 1959 Draper Report in the opening sequence. He also cited the figures from Coale and Hoover 1958.
3 The Shaker hymn "Simple Gifts" was played as the report opening and closing credits ran. The lyrics begin, "'Tis the gift to be simple, 'tis the gift to be free, 'tis the gift to come down where we ought to be." The nineteenth-century hymn was popularized when Aaron Copeland included it in his 1944 *Appalachian Spring* and his 1950 collection of American folk songs. Both symbolize traditional rural life.
4 Finkle and Crane 1975, 91, 89–90.
5 "Declaration of Population" 1967. The signatories to the resolution included Australia, Barbados, Colombia, Denmark, the Dominican Republic, Finland, the National Liberation Council of Ghana, India, Indonesia, Iran, Japan, Jordan, Korea, Malaysia, Morocco, Nepal, the Netherlands, New Zealand, Norway, Pakistan, the Philippines, Singapore, Sweden, Thailand, Trinidad and Tobago, Tunisia, UAR (Egypt), the United Kingdom, the United States, and Yugoslavia.

6 Berelson 1969, 533, notes that there was agreement that population growth should be reduced, but that there was disagreement as to means. Ehrlich 1968 and Kingsley Davis 1967 supported zero growth, which requires fertility and mortality rates to be balanced at the replacement level.
7 Notestein 1969b, 35–36.
8 Finkle and Crane 1975, 87; Mauldin et al. 1974, 358; Connelly 2008, 299.
9 Berelson 1975, 130–131. Mauldin et al. 1974, 395, notes that the "rule of thumb" estimate is that a 30% rate of contraceptive use by reproductive-age women will produce a 30 per 1,000 crude birth rate.
10 Mauldin et al. 1974, 392, 394, 395.
11 Berelson et al. 1966, viii. Dunlop interview, 2004, 18, notes, however, that by 1974 the idea of population control was stale and the vitality belonged to the opposition.
12 See Prashad 2007, 132, 189. See also Finkle and Crane 1975, 99, 101–104; McIntosh and Finkle 1995, 232.
13 World Population Conference 1975, 166, 168, 171; Mauldin et al. 1974, 372.
14 Mauldin et al. 1974, 376. In addition to non-aligned nations, the coalition of actors that defeated the demographic targets included the longstanding opponents of population control among Communist and Catholic nations and NGOs. The Vatican, of course, opposed contraception on moral grounds. Communist nations objected to the idea of population crisis based on Marx's critique of Malthus: poverty resulted from the inequitable distribution of socially produced wealth within capitalism. See also Connelly 2008, 304, 310–316; Critchlow 1999, 181.
15 Mauldin et al. 1974, 379.
16 For this reason, Connelly 2008, 316, calls the Bucharest conference "the Waterloo" of population control. Funding for family planning programs declined during the remainder of the decade, as did all foreign aid. Yet an additional $2 billion was spent in developing nations in 1980. This figure was considerably less than what population control advocates recommended. Draper set the needed figure at $3 billion in 1974. See Connelly 2008, 327, 299. The Ford Foundation decreased its commitment to population and shifted to supporting programs for women. See Critchlow 1999, 187–188. Demographic fortunes fell more substantially as a result of the 1984 UN Population Conference. At that conference, conservative religious forces, with the backing of the United States, now under the leadership of Ronald Reagan, declared that population growth was a neutral factor in economic development, and thus did not warrant focused attention. At that point, funding began to melt away. See Hodgson and Watkins 1997, 493–495.
17 Finkle and Crane 1975, 105; see also Connelly 2008, 301–302.
18 World Population Conference 1975, 164. See also Finkle and Crane 1975, 91–94, for an account of the influence of the non-aligned movement and the New International Economic Order on the conference proceedings. They note the importance of Algeria, an OPEC nation, in this rejection. In

the summer of 1974, OPEC, an alliance of oil-producing nations, successfully challenged the international power of developed nations when it raised world oil prices.

19 World Population Conference 1975, 165, 168, 171. The plan, 167, stated that human rights must be protected, but provided no guidance on how that might be assured. Hodgson and Watkins 1997, 490, note that family planning programs proliferated; by 1991, 69 nations had population control policies (495).

20 Connelly 2008, 296, 317–326.

21 Germain interview, 2003, 64.

22 This was the first of many conferences at which Joan Dunlop and Adrienne Germain played significant roles. Dunlop interview, 2004, 8–9, notes the importance of Panamanian demographer Carmen Miró of the IUSSP in helping to bring the issue forward. Feminists were also anticipating the first UN International Women's Conference, slated for the following year, which the UN declared Women's Year. The resolution establishing Women's Year and calling for the conference was approved in 1972, the same year as the resolutions for Population Year and a population conference.

23 World Population Conference 1975, 167; Mauldin et al. 1974, 364, 375, 378; Berelson 1975, 118, 126ff, 134–135.

24 Connelly 2008, 315.

25 Rockefeller 1978, 515–516. See also Germain interview, 2003, 48–50; Dunlop 2004, 9–15, on the speech-writing process and its impact.

26 Rockefeller did not speak before the formal conference delegates. He spoke at one of five lectures organized by the IUSSP for the Tribune. See Rockefeller 1978 [1974], 509. Berelson, then council president, and Notestein were especially disconcerted by Rockefeller's shift and tried to dissuade him. See Critchlow 1999, 179–183. Notestein criticized the speech in his obituary of Rockefeller. See Notestein 1978, 506. See also Dunlop interview, 2004, 14–16; Germain interview, 2003, 49–54.

27 Connelly 2008, 315.

28 Joan Dunlop, who served as an adviser to Rockefeller beginning in 1973, persuaded him that the council's approach was detrimental to international relations and to individuals, and urged him to refocus efforts on women's reproductive rights. Thereafter he did focus on abortion rights until his untimely death in 1978. The council was in turmoil for a number of years as it struggled to select a new president. See Dunlop interview, 2004, 17–20; Connelly 2008, 329–330.

29 Mauldin et al. 1974, 373.

30 Ibid.; Berelson 1975. See also McIntosh and Finkle 1995, 234.

31 Germain interview, 2003, 62–63, 65, 149–152.

32 Connelly 2008, 316. Dunlop is referring to the aftermath of Bucharest. See also Germain interview, 2003, 21, 24. While Berelson, Notestein, and others blamed Dunlop for changing Rockefeller's mind, she recounts that

when she was hired, he said he thought something was "wrong with the population field" (Dunlop interview, 2004, 4–5; Germain interview, 2003, 44).

33 See Westoff 1978, 1981, 1988a, 1988b, 1992; Westoff and Pebley 1981; Nortman 1982; Bongaarts 1990, 1991. See also Robbins 1973 for an early definition by family planners.

34 Westoff 1994.

35 Studies of unmet need still proliferate. A quick journal article search on the keywords *unmet need* and *contraception* yields 1,000 entries. On post-Cairo metrics of unmet need, see Halfon 2007.

36 Hodgson and Watkins, 1997, 498–499. Germain and Ordway 1989 and Bruce 1990 are examples.

37 The Cairo conference was the largest of its kind, with an estimated 20,000 to 25,000 people attending, representing 180 national delegations, 2,000 NGOs, and some 3,000 journalists and media representatives. Johnson 1995, 88.

38 "Women's Declaration" 1994, 32. Dunlop interview, 2004, 101, 128–129, notes that the strength of the IWHC network was its diversity and size. Alliance members went to each of the preparatory conferences, as well as the final conference, to monitor the proceedings and to lobby for the inclusion of language consistent with the Women's Declaration. The IWHC raised $1 million to support travel by alliance members to the Cairo conference and pre-meetings. See Germain interview, 2003, 149–154, and Dunlop interview, 2004, 93–100, 109, for a detailed description.

39 Hodgson and Watkins 1997, 499. On the influence of the IWHC, see also McIntosh and Finkle 1995, 237–238, 242.

40 "Women's Declaration" 1994, 32.

41 Gita Sen, Germain, and Chen 1994, 4.

42 "Women's Declaration" 1994, 32. In addition, the declaration charges, existing policies focus too narrowly on managing women's bodies. Yet the behavior of their male partners greatly affects women's concrete ability to make reproductive decisions, and, therefore, policies should empower men to take responsibility for fertility control as well.

43 United Nations 2014 [1994], 12, 59.

44 McIntosh and Finkle 1995, 227, 249, 253, n. 6, indicate that this is the first UN population conference in which reproductive rights were specifically identified as belonging to women, as opposed to men, families, or abstract individuals. See Dunlop interview, 2004, 100–112, 118–134, and Germain interview, 2003, 157–162, 178, for a detailed account of efforts at the conference made by the IWHC network to protect that language. They both note that the actions of members from the Global South contributed substantially to the IWHC network's successes at the conference.

45 United Nations 2014 [1994], 30, 72. See Germain interview, 2003, 162–167, on the failure to win endorsement of women's sexual rights.

46 United Nations 2014 [1994], 49, 53, 60, 63, 72–73, 76.
47 Ibid., 64, 68, 242.
48 Ibid., 64, 66–69, 83–90.
49 Hodgson and Watkins 1997 is an interesting bit of disciplinary boundary work. The authors rehearse the history of population control, emphasizing a longstanding alliance between feminists and what it labels neo-Malthusians. They define neo-Malthusianism as a social movement separate from demography but composed of demographers, foundations, and family planners. On this basis, they assert that Cairo achieved a "common ground agenda" between the two groups premised on demographic convictions. But they conclude that the alliance is fragile because the common ground is too narrow and depends on the rejection of some demographic claims. They also suggest, 494–495, that feminists dominated in Cairo because neo-Malthusianism had waned, and they express hope, 511, for the resurgence of neo-Malthusianism with the growing importance of demographers in the Global South.
50 See McIntosh and Finkle 1995, 227, 235, 242, 249–250, whose account of the run-up to Cairo begins with the first UN population conference in 1954, which was entirely the purview of demographers. See also Hodgson and Watkins 1997, 506.
51 McIntosh and Finkle 1995, 230; Hodgson and Watkins 1997, 499–500.
52 McIntosh and Finkle 1995, 227; Westoff 1994.
53 McIntosh and Finkle 1995, 230.
54 Ibid., 234, 251.
55 Westoff 1994.
56 McIntosh and Finkle 1995, 252.
57 Ibid., 235, 230.
58 Ibid., 252, 242; Westoff 1994. See also Hodgson and Watkins 1997, 498–500.
59 See Westoff 1994 and Hodgson and Watkins 1997, 507, on the shaky empirical basis of the feminist claim that empowering women will lead to reduced fertility.
60 United Nations 2014 [1994], 46.
61 Ibid.
62 Gita Sen, Germain, and Chen 1994, 3. See also Murphy 2012, 142–145, who notes as well that the Cairo Programme of Action left the neoliberal development perspective untouched.
63 Germain interview, 2003, 144.
64 Hodgson and Watkins 1997, 500–501, make this point as well, although in reading it as a positive connection for neo-Malthusianism, I think they overstate the case of an alliance between demographic neo-Malthusians and feminists.
65 Connelly 2008, 278, 311–312. Connelly takes the phrase *system without a brain* from Symonds and Carder 1978, 192, whose Population Council–funded book on population and the UN took the phrase from a 1969 report

that used it to describe all the various UN development programs other than population.

66 Connelly 2008, 277. Where Connelly 2008 attends to the politics of population control, Bashford 2014, 4–5, 354, 361, eschews an analysis of power in her intellectual history of global population discourse, focusing only on the knowledge portion of the knowledge-power nexus by which the crisis problematization was configured. Her account broadens the range of perspectives on population matters, particularly "the colonial history of world population" (361). But by assuming an apolitical analytic stance, she lets population science off the hook.

67 Is it any wonder, then, that midcentury contraceptive research focused on containing ova, not sperm? See Takeshita 2011 on the IUD and Marks 2001 on the pill.

68 For a critical feminist demography of fertility change, much of which is grounded in anticolonial and feminist anthropology, see, for example, Gita Sen, Germain, and Chen 1994; Greenhalgh 2005, 1995; Greenhalgh and Li 1995; Silliman and King 1999; Riley and McCarthy 2003; Szreter et al. 2004; Browner and Sargent 2011.

69 Germain interview, 2003, 188–195, notes in particular the rise of quantitative epidemiological models. See also Murphy 2012. Halfon 2007 provides a focused critique of unmet need.

70 See Moya 2001, 477–478, on this feature of postpositivist realism.

References

Archival Sources

Dunlop, Joan. Interview by Rebecca Sharpless. Transcript of audio recording, April 14–15, 2004. Population and Reproductive Health Oral History Project, Sophia Smith Collection, Smith College, Northampton, MA.
Ferguson, Frances Hand. Interview by James Reed. Transcript of audio recording, June 3, 1974. Schlesinger-Rockefeller Oral History Project, Schlesinger Library, Radcliffe College, Cambridge, MA.
Germain, Adrienne. Interview by Rebecca Sharpless. Transcript of audio recording, June 19–20 and September 25, 2003. Population and Reproductive Health Oral History Project, Sophia Smith Collection, Smith College, Northampton, MA.
John D. Rockefeller 3rd Papers, Rockefeller Archive Center, Sleepy Hollow, NY.
Margaret Sanger Papers, Library of Congress, Washington, DC.
Office of the Messrs. Rockefeller Records, Rockefeller Archive Center, Sleepy Hollow, NY.
Piotrow, Phyllis Tilson. Interview by Rebecca Sharpless. Transcript of audio recording, September 16, 2002. Population and Reproductive Health Oral History Project, Sophia Smith Collection, Smith College, Northampton, MA.
Population Council Records, Rockefeller Archive Center, Sleepy Hollow, NY.

Books and Articles

Ahluwalia, Sanjam. 2008. *Reproductive Restraints: Birth Control in India, 1877–1947*. Urbana: University of Illinois Press.
———. 2004. "Demographic Rhetoric and Sexual Surveillance: Indian Middle-Class Advocates of Birth Control, 1920s-1940s." In *Confronting the Body: The Politics of Physicality in Colonial and Post-Colonial India*, ed. James H. Mills and Satadru Sen, 183–200. London: Anthem Press.
Ahmed, Sara. 2010. *The Promise of Happiness*. Durham: Duke University Press.

———. 2007/2008. "Multiculturalism and the Promise of Happiness." *New Formations* 63 (Winter): 121–137.
———. 2004a. "Affective Economies." *Social Text* 79, 22, no. 2: 117–139.
———. 2004b. *The Cultural Politics of Emotion.* New York: Routledge.
Alborn, Timothy. 1999. "Age and Empire in the Indian Census, 1971–1931." *Journal of Interdisciplinary History* 30, no. 1 (Summer): 61–89.
Anderson, Margo. 1988. *The American Census.* New Haven: Yale University Press.
Anderson, Benedict. 1991. *Imagined Communities.* Rev ed. London: Verso.
Appadurai, Arjun. 1996. "Number in the Colonial Imagination." In *Modernity at Large: Cultural Dimensions of Globalization*, 114–135. Minneapolis: University of Minnesota Press.
Arnold, David. 2006. "Official Attitudes to Population, Birth Control and Reproductive Health in India, 1921–1946." In *Reproductive Health in India: History, Politics, Controversies,* ed. Sarah Hodges, 22–50. New Delhi: Orient Longman.
Aryee, Anna. 2006. "Gandhi and Mrs. Sanger Debate Birth Control: Comment." In *Reproductive Health in India: History, Politics, Controversies,* ed. Sarah Hodges, 227–234. New Delhi: Orient Longman.
Balfour, Marshall, Roger Evans, Frank Notestein, and Irene Taeuber. 1950. *Public Health and Demography in the Far East.* New York: Rockefeller Foundation.
Balibar, Étienne, and Immanuel Wallerstein. 1991. *Race, Nation, Class.* New York: Verso.
Bandarage, Asoka. 1997. *Women, Population and Global Crisis: A Political-Economic Analysis.* London: Zed Books.
Banton, Michael. 2002. *The International Politics of Race.* London: Polity Press.
Bashford, Alison. 2014. *Global Population: History, Geo-Politics, and Life on Earth.* New York: Columbia University Press.
———. 2012. "Anticolonial Climates: Physiology, Ecology, and Global Population, 1920s-1950s. *Bulletin of the History of Medicine* 86, no. 4 (Winter): 596–626.
———. 2008. "Population, Geopolitics, and International Organizations in the Mid-Twentieth Century." *Journal of World History* 19, no. 3 (September): 327–348.
———. 2007. "Nation, Empire, Globe: The Spaces of Population Debate in the Interwar Years." *Comparative Studies in Society and History* 49, no. 1 (January): 170–201.
Berelson, Bernard. 1975. "World Population Plan of Action: Where Now?" *Population and Development Review* 1, no. 1 (September): 115–146.
———. 1969. "Beyond Family Planning." *Science* 163, no. 3867 (February): 533–543.
Berelson, Bernard, Richard K. Anderson, Oscar Harkavy, John Maier, W. Parker Mauldin, and Sheldon J. Segal. 1966. *Family Planning and Population Programs: A Review of World Developments.* Chicago: University of Chicago Press.
———. 1964. "National Family Planning Programs: A Guide." *Studies in Family Planning* 1, no. 5, Supplement (December): 1–12.

Bergeron, Suzanne. 2006. *Fragments of Development: Nation, Gender, and the Space of Modernity.* Ann Arbor: University of Michigan Press.
Berlant, Lauren. 2008. *The Female Complaint: The Unfinished Business of Sentimentality in American Culture.* Durham: Duke University Press.
Bhatia, Rajani, and Ashwini Tambe. 2014. "Raising the Age of Marriage in 1970s India: Demographers, Despots, and Feminists." *Women's Studies International Forum* 44: 89–100.
Bhore, Joseph. 1945. *Health and Development Committee Report.* Vol. 2. Dehli: Government of India.
"Birth Control in India: Mrs. How-Martyn on Her Tour." *Marriage Hygiene* 3 (February 1937): 214–242.
Bongaarts, John. 1991. "The KAP-Gap and the Unmet Need for Contraception." *Population and Development Review* 17, no. 2 (June): 293–313.
———. 1990. "The Measurement of Unwanted Fertility." *Population and Development Review* 16, no. 3 (September): 487–506.
Bose, Subhas Chandra. 1997. "The Haripura Address." In *The Essential Writings of Netaji Subhas Chandra Bose,* ed. Sisir K. Bose and Sugata Bose, 197–219. Delhi: Oxford University Press.
Breslau, Daniel. 2007. "The American Spencerians: Theorizing a New Science." In *Sociology in America: A History,* ed. Craig Calhoun, 39–62. Chicago: University of Chicago Press.
Briggs, Laura. 2002. *Reproducing Empire: Race, Sex, Science, and U.S. Imperialism in Puerto Rico.* Berkeley: University of California Press.
Britt, Elizabeth. 2000. "Medical Insurance as Bio-Power: Law and the Normalization of (In)fertility." In *Body Talk: Rhetoric, Technology, Reproduction,* ed. Mary Lay, Laura Gurak, Clare Gravon, and Cynthia Myntti, 207–225 Madison: University of Wisconsin Press.
Browner, Carole, and Carolyn Sargent. 2011. *Reproduction, Globalization and the State: New Theoretical and Ethnographic Perspectives.* Durham: Duke University Press.
Bruce, Judith. 1990. "Fundamentals of the Quality of Care: A Simple Framework." *Studies in Family Planning* 21, no. 2 (March-April): 61–91.
Burchell, Graham, Colin Gordon, and Peter Miller, eds. 1991. *The Foucault Effect: Studies in Governmentality.* Chicago: University of Chicago Press.
Butler, Judith. 1993. *Bodies That Matter: On the Discursive Limits of "Sex."* New York: Routledge.
Caldwell, John. 1998. "Malthus and the Less Developed World: The Pivotal Role of India." *Population and Development Review* 24, no. 4 (December): 675–696.
Caldwell, John, and Pat Caldwell. 1986. *Limiting Population Growth and the Ford Foundation Contribution.* London: Frances Pinter Press.
Calhoun, Craig, ed. 2007. *Sociology in America: A History.* Chicago: University of Chicago Press.
Calhoun, Craig, and Jonathan VanAntwerpen. 2007. "Orthodoxy, Heterodoxy, and Hierarchy: 'Mainstream' Sociology and Its Challengers." In *Sociology*

in America: A History, ed. Craig Calhoun, 367–410. Chicago: University of Chicago Press.

Camic, Charles. 2007. "On the Edge: Sociology during the Great Depression and the New Deal." In *Sociology in America: A History*, ed. Craig Calhoun, 225–280. Chicago: University of Chicago Press.

Campbell, Nancy. 2000. *Using Women: Gender, Drug Policy, and Social Justice.* New York: Routledge.

Carrigan, Tim, Bob Connell, and John Lee. 1985. "Towards a New Sociology of Masculinity." *Theory and Society* 14, no. 5: 551–604.

Carr-Saunders, Alexander. 1936. *World Population: Past Growth and Present Trends.* Oxford: Clarendon Press.

———. 1925. *Population.* London: Oxford University Press.

———. 1922. *A Population Problem: A Study in Human Evolution.* London: Oxford University Press.

Casper, Monica, and Lisa Jean Moore. 2009. *Missing Bodies: The Politics of Visibility.* New York: New York University Press.

Chand, Gyan. 1944. *The Problem of Population.* Oxford Pamphlets on Indian Affairs, no. 19. London: Humphrey Milford, University Press.

———. 1939. *India's Teeming Millions*. New York: W. W. Norton.

Chandrasekhar, Sripati. 1967. *India's Population: Facts, Problem and Policy*. Delhi: Meenakshi Prakashan.

———. 1961 [1955]. *Population and Planned Parenthood in India.* 2nd ed. London: George Allen and Unwin.

———. 1959. *Infant Mortality in India, 1901–1955*. London: George Allen and Unwin.

———. 1944. "Recent Social Science Literature in India." *Annals of the American Academy of Political and Social Science* 233, India Speaking (May): 208–217.

Chatterjee, Nilanjana, and Nancy E. Riley. 2001. "Planning an Indian Modernity: The Gendered Politics of Fertility Control." *Signs: Journal of Women in Culture and Society* 26, no. 3 (Spring): 811–845.

Chesler, Ellen. 1992. *Woman of Valor: Margaret Sanger and the Birth Control Movement in America.* New York: Simon and Schuster.

Clarke, Adele. 1998. *Disciplining Reproduction: Modernity, American Life Sciences, and the Problems of Sex.* Berkeley: University of California Press.

Coale, Ansley. 1983. "Frank Notestein, 1902–1983." *Population Index* 49, no. 1 (Spring): 3–12.

Coale, Ansley, and Edgar M. Hoover. 1958. *Population Growth and Economic Development in Low-Income Countries: A Case Study of India's Prospects.* Princeton: Princeton University Press.

Cohn, Bernard. 1996. *Colonialism and Its Forms of Knowledge: The British in India.* Princeton: Princeton University Press.

———. 1987. "The Census, Social Structure, and Objectification in South Asia." In *An Anthropologist among the Historians and Other Essays,* 224–254. New York: Oxford University Press.

Collins, Patricia Hill. 1999. "Producing the Mothers of the Nation: Race, Class,

and Contemporary US Population Policies." In *Women, Citizenship and Difference*, ed. Nira Yuval-Davis and Pnina Werbner, 118–129. London: Zed Books.
Connell, Raewyn. 2005a. "Globalization, Imperialism, and Masculinities." In *Handbook of Studies on Men and Masculinities*, ed. Michael S. Kimmel, Jeff Hearn, and R. W. Connell, 71–89. Thousand Oaks, CA: Sage.
———. 2005b. *Masculinities*. Berkeley: University of California Press.
———. 1987. *Gender and Power*. Palo Alto: Stanford University Press.
Connell, Raewyn, and James W. Messerschmidt. 2005. "Hegemonic Masculinity: Rethinking the Concept." *Gender and Society* 19, no. 6 (December): 829–859, 832.
Connelly, Matthew. 2008. *Fatal Misconception: The Struggle to Control World Population*. Cambridge, MA: Harvard University Press.
———. 2006. "To Inherit the Earth: Imagining World Population, from the Yellow Peril to the Population Bomb." *Journal of Global History* 1, no. 1: 299–319.
Cooper, Brian. 2003. "Social Classifications, Social Statistics, and the 'Facts' of 'Difference' in Economics." In *Toward a Feminist Philosophy of Economics*, ed. Drucilla Barker and Edith Kuiper, 161–179. New York: Routledge.
Cooper, Frederick, and Ann Laura Stoler. 1997. *Tensions of Empire: Colonial Cultures in a Bourgeois World*. Berkeley: University of California Press.
Critchlow, Donald. 1999. *Intended Consequences: Birth Control, Abortion, and the Federal Government in Modern America*. New York: Oxford University Press.
Curtis, Bruce. 2001. *The Politics of Population: State Formation, Statistics, and the Census of Canada, 1840–1875*. Toronto: University of Toronto Press.
Danby, Colin. 2012. "Post War Norm." *Rethinking Marxism: A Journal of Economics, Culture, and Society* 24, no. 4: 499–515.
Davenport, Randi. 1995. "Thomas Malthus and Maternal Bodies Politic: Gender, Race, and Empire." *Women's History Review* 4, no. 4: 415–439.
Davis, Kathy. 2007. *The Making of Our Bodies, Ourselves: How Feminisms Travel across Borders*. Durham: Duke University Press.
Davis, Kingsley. 1967. "Population Policy: Will Current Programs Succeed?" *Science* 158, no. 3802 (November): 730–739.
———. 1963. "The Theory of Change and Response in Modern Demographic History." *Population Index* 29, no. 4 (October): 345–366.
———. 1959. "The Other Scare: Too Many People." *New York Times*, March 15.
———. 1957. "Analysis of the Population Explosion." *New York Times*, September 22.
———. 1956. "The Population Specter: Rapidly Declining Death Rate in Densely Populated Countries—The Amazing Decline of Mortality in Undeveloped Areas." *American Economic Review* 46, no. 2 (May): 305–318.
———. 1955. "Social and Demographic Aspects of Economic Development in India." In *Economic Growth: Brazil, India and Japan*, eds. Simon Kuznets et al., 263–315. Durham: Duke University Press, 1955.
———. 1951a. "Population and the Further Spread of Industrial Society." *American Philosophical Society* 95, no. 1 (February 13): 8–19.

———. 1951b. *The Population of India and Pakistan.* Princeton: Princeton University Press.
———. 1950a. "The Economic Demography of India and Pakistan." In *South Asia in the World Today,* ed. Phillip Talbot, 86–107. Chicago: University of Chicago Press.
———. 1950b. "Population and Change in Backward Areas." *Columbia Journal of International Affairs* 4, no. 2 (Spring): 43–49.
———. 1948. *Human Society*. New York: Macmillan.
———. 1946. "Human Fertility in India." *American Journal of Sociology* 52, no. 3 (November): 243–254.
———. 1945. "The World Demographic Transition." *Annals of the American Academy of Political and Social Science* 237 (January): 1–11.
———. 1944. "Demographic Fact and Policy in India." *Milbank Memorial Fund Quarterly* 22, no. 3 (July): 256–278.
———. 1943. "The Population of India." *Far East Survey* 12, no. 8 (April 19): 76–79.
———. 1937. "Reproductive Institutions and the Pressure of Population." *Sociological Review* 29, no. 3 (July): 289–306.
Davis, Kingsley, and Judith Blake. 1956. "Social Structure and Fertility: An Analytic Framework." *Economic Development and Cultural Change* 4, no. 3 (April): 211–235.
Davis, Mike. 2001. *Late Victorian Holocausts: El Niño Famines and the Making of the Third World.* New York: Verso.
"Declaration of Population." 1967. *Studies in Family Planning* 1, no. 16 (January): 1.
Demeny, Paul. 1988. "Social Science and Population Policy." *Population and Development Review* 14, no. 3 (September): 451–479.
Demetriou, Demetrakis Z. 2001. "Connell's Concept of Hegemonic Masculinity: A Critique." *Theory and Society* 30, no. 3 (June): 337–361.
Desrosières, Alain. 1998. *The Politics of Large Numbers: A History of Statistical Reasoning.* Cambridge: Harvard University Press.
———. 1990. "How to Make Things Which Hold Together: Social Science, Statistics and the State." In *Discourses on Society,* ed. P. Wagner, Björn Wittrock, and R. Whitley, 195–218. Amsterdam: Kluwer.
Dublin, Louis, and Alfred Lotka. 1925. "On the True Rate of Natural Increase." *Journal of the American Statistical Association* 20, no. 151 (September): 305–339.
Duden, Barbara. 1992. "Population." In *The Development Dictionary,* ed. Wolfgang Sachs, 146–157. London: Zed Books.
Ehrlich, Paul. 1968. *The Population Bomb: Population Control or Race to Oblivion.* New York: Sierra Club/Ballantine.
Escobar, Arturo. 1995. *Encountering Development: The Making and Unmaking of the Third World.* Princeton: Princeton University Press.
Fairchild, Henry Pratt. 1934. "Organization for Research in Population." *Human Biology* 6, no. 1 (February): 223–239.

Ferree, Myra, Shamus Rahman Khan, and Shauna A. Morimoto. 2007. "Assessing the Feminist Revolution: The Presence and Absence of Gender in Theory and Practice." In *Sociology in America: A History*, ed. Craig Calhoun, 438–479. Chicago: University of Chicago Press.
Finkle, Jason, and Barbara Crane. 1975. "The Politics of Bucharest: Population, Development, and the New International Economic Order." *Population and Development Review* 1, no. 1: 87–114.
Florence, Lella. 1930. *Birth Control on Trial*. London: George Allen and Unwin.
Foucault, Michel. 2007. *Security, Territory, Population*. New York: Palgrave Macmillan.
———. 1991. "Governmentality." In *The Foucault Effect: Studies in Governmentality*, ed. Graham Burchell, Colin Gordon, and Peter Miller, 87–104. Chicago: University of Chicago Press.
———. 1979. *Discipline and Punish*. New York: Vintage.
———. 1978. *A History of Sexuality. Volume One: An Introduction*. New York: Vintage.
Gardiner, Judith Kegan. 2002. "Introduction." In *Masculinity Studies and Feminist Theory: New Directions*, ed. Judith Kegan Gardiner, 1–29. New York: Columbia University Press.
Germain, Adrienne, and Jane Ordway. 1989. *Population Control and Women's Health: Balancing the Scales*. New York: International Women's Health Coalition.
Gieryn, Thomas. 1999. *Cultural Boundaries of Science*. Chicago: University of Chicago Press.
———. 1995. "Boundaries of Science." In *Handbook of Science and Technology Studies*, ed. Sheila Jasanoff et al., 393–443. Thousand Oaks, CA: Sage.
Goldberg, David Theo. 1993. *Racist Culture: Philosophy and the Politics of Meaning*. Oxford: Blackwell.
Gordon, Avery. 2008. *Ghostly Matters: Haunting and the Sociological Imagination*. Minneapolis: University of Minnesota Press.
Gordon, Linda. 2002. *The Moral Property of Women: A History of Birth Control Politics in America*. Urbana: University of Illinois Press.
———. 1990. *Woman's Body, Woman's Right: Birth Control in America*. 2nd ed. New York: Penguin.
———. 1976. *Woman's Body, Woman's Right: Birth Control in America*. New York: Penguin.
Greene, Ronald Walter. 1999. *Malthusian Worlds: U.S. Leadership and the Governing of the Population Crisis*. Boulder, CO: Westview.
Greenhalgh, Susan. 2008. *Just One Child: Science and Policy in Deng's China*. Berkeley: University of California Press.
———. 2005. "Globalization and Population Governance in China." In *Global Assemblages: Technology, Politics, and Ethic as Anthropological Problems*, ed. Aihwa Ong and Stephen J. Collier, 354–372. London: Blackwell.
———. 1996. "The Social Construction of Population Science: An Intellectual, Institutional and Political History of Twentieth Century Demography." *Comparative Studies of Society and History* 5, no. 1 (January): 26–66.

———. 1995. "Anthropology Theorizes Reproduction." In *Situating Fertility: Anthropology and Demographic Inquiry,* ed. Susan Greenhalgh, 3–28. Cambridge: Cambridge University Press.

Greenhalgh, Susan, and Jiali Li. 1995. "Engendering Reproductive Policy and Practice in Peasant China: For a Feminist Demography of Reproduction." *Signs: Journal of Women in Culture and Society* 20, no. 3 (Spring): 601–641.

Guilmoto, Christophe. 2005. "Fertility Decline in India: Maps, Models, and Hypotheses." In *Fertility Transition in South India,* ed. Christophe Guilmoto and S. Iruday Rajan, 385–435. New Delhi: Sage.

Hacking, Ian. 1992. "Statistical Language, Statistical Truth and Statistical Reason: The Self-Authentication of a Style of Scientific Reason." In *The Social Dimension of Science,* ed. Ernan McMullin, 130–157. South Bend, IN: University of Notre Dame Press.

———. 1990. *The Taming of Chance.* Cambridge: Cambridge University Press.

———. 1986. "Making up People." In *Reconstructing Individualism,* ed. Thomas Heller, Morton Sosna, and David Wellbery, 222–236. Stanford: Stanford University Press.

———. 1982. "Biopower and the Avalanche of Printed Numbers." *Humanities in Society* 5, no. 3/4: 279–295.

Halfon, Saul. 2007. *The Cairo Consensus: Demographic Surveys, Women's Empowerment, and Regime Change in Population Policy.* Lanham, MD: Lexington Books.

Hall, Stuart, Chas Critcher, Tony Jefferson, John Clarke, and Brian Roberts. 1978. *Policing the Crisis: Mugging, the State, and Law and Order.* London: Macmillan.

Haraway, Donna. 2000. *How like a Leaf: An Interview with Thyrza Nichols Goodeve.* New York: Routledge.

———. 1997. "Race: Universal Donors in a Vampire Culture: It's All in the Family. Biological Kinship Categories in the Twentieth Century United States." In *Modest Witness@Second Millennium,* ed. Donna Haraway, 213–265. New York: Routledge.

Hartmann, Betsy. 1995. *Reproductive Rights and Wrongs: The Global Politics of Population Control.* Boston: South End Press.

Hauser, Philip, and Otis Dudley Duncan. 1959. *The Study of Population: An Inventory and Appraisal.* Chicago: University of Chicago Press.

Heer, David. 2004. *Kingsley Davis: A Biography and Selections from His Writings.* New Brunswick: Transaction.

Himes, Norman. 1936. *Medical History of Contraception.* Baltimore: Williams and Wilkins.

Hodges, Sarah. 2008. *Contraception, Colonialism and Commerce: Birth Control in South India, 1920–1940.* Burlington, VT: Ashgate.

———. 2006a. "Indian Eugenics in the Age of Reform." In *Reproductive Health in India: History, Politics, Controversies,* ed. Sarah Hodges, 115–138. New Delhi: Orient Longman.

———. 2006b. *Reproductive Health in India: History, Politics, Controversies.* New Delhi: Orient Longman.

———. 2004. "Governmentality, Population and Reproductive Family in Modern India." *Economic and Political Weekly* 39, no. 11 (March): 1157–1163.

Hodgson, Dennis. 1991. "The Ideological Origins of the Population Association of America." *Population and Development Review* 17, no. 1: 1–34.

———. 1988. "Orthodoxy and Revisionism in American Demography." *Population and Development Review* 14, no. 4: 541–569.

———. 1983. "Demography as Social Science and Policy Science." *Population and Development Review* 9, no. 1: 1–34.

Hodgson, Dennis, and Susan C. Watkins. 1997. "Feminists and Neo-Malthusians: Past and Present Alliances." *Population and Development Review* 23, no. 3: 469–523.

Hooper, Charlotte. 2001. *Manly States: Masculinities, International Relations, and Gender Politics.* New York: Columbia University Press.

Horn, David. 1994. *Social Bodies: Science, Reproduction, and Italian Modernity.* Princeton: Princeton University Press.

Howard, June. 1999. "What Is Sentimentality?" *American Literary History* 11, no. 1 (Spring): 63–81.

Igo, Sarah. 2007. *The Averaged American.* Cambridge: Harvard University Press.

James, Patricia. 1979. *Population Malthus.* London: Routledge.

Johnson, Stanley. 1995. *The Politics of Population: Cairo 1994.* London: Earthscan.

Jordanova, Ludmilla. 1989. *Sexual Visions: Images of Gender Science and Medicine between the Eighteenth and Twentieth Centuries.* Madison: University of Wisconsin Press.

Joshi, P. C. 1986. "Founders of the Lucknow School and Their Legacy: Radhakamal Mukerjee and D. P. Mukerji: Some Reflections." *Economic and Political Weekly* 21, no. 33 (August 16): 1455–1469.

Kevles, Daniel. 1985. *In the Name of Eugenics.* Berkeley: University of California Press.

Kirk, Dudley. 1996. "Demographic Transition Theory." *Population Studies* 50, no. 3 (November): 361–387.

———. 1960. "Some Reflections on American Demography in the Nineteen Sixties." *Population Index* 26, no. 4 (October): 305–310.

———. 1955. "Dynamics of Human Populations." *Eugenics Quarterly* 2: 18–25.

———. 1944. "Population Changes and the Postwar World." *American Sociological Review* 9, no. 1 (February): 28–35.

Kiser, Clyde. 1981. "The Role of the Milbank Memorial Fund in the Early History of the Association." *Population Index* 47, no. 3: 490–494.

———. 1971. "The Work of the Milbank Memorial Fund in Population since 1928." In *Forty Years of Research in Human Fertility,* ed. Clyde Kiser, 15–62. New York: Milbank Memorial Fund.

———. 1953. "The Population Association Comes of Age." *Eugenical News* 38 (December): 107–111.

Kopp, Marie. 1933. *Birth Control in Practice.* New York: Robert McBride.

Ladd-Taylor, Molly. 2001. "Eugenics, Sterilization and Modern Marriage in

the USA: The Strange Career of Paul Popenoe." *Gender and History* 13, no. 2 (August): 298–327.
Laslett, Barbara. 2007. "Feminist Sociology in the Twentieth Century United States: Life Stories in Historical Context." In *Sociology in America: A History,* ed. Craig Calhoun, 480–502. Chicago: University of Chicago Press.
Latham, Michael. 2000. *Modernization as Ideology: American Social Science and "Nation Building" in the Kennedy Era.* Chapel Hill: University of North Carolina Press.
Latour, Bruno. 1987. *Science In Action: How to Follow Scientists and Engineers through Society.* Cambridge: Harvard University Press.
Leavitt, Judith Walzer, 1986. *Brought to Bed: Childbearing in America, 1750–1950.* New York: Oxford University Press.
Ledbetter, Rosanna. 1984. "Thirty Years of Family Planning in India." *Asian Survey* 24, no. 7 (July): 736–758.
Lengermann, Patricia Madoo, and Gillian Niebrugge. 2007. "Thrice Told: Narratives of Sociology's Relation to Social Work." In *Sociology in America: A History,* ed. Craig Calhoun, 63–114. Chicago: University of Chicago Press.
Lengermann, Patricia Madoo, and Gillian Niebrugge-Brantley. 1998. *The Women Founders: Sociology and Social Theory, 1830–1930.* New York: McGraw-Hill.
Lipsitz, George. 1988. *Ivory Perry: A Life in the Struggle.* Philadelphia: Temple University Press.
Lorimer, Frank. 1981. "How Demographers Saved the Association." *Population Index* 47, no. 3: 488–490.
———. 1971. "The Role of the International Union for the Scientific Study of Population." In *Forty Years of Research in Human Fertility,* ed. Clyde Kiser, 86–97. New York: Milbank Memorial Fund.
———. 1951. "Dynamics of Age Structure in a Population with Initially High Fertility and Mortality." *Population Bulletin,* no. 1 (December 1951): 31–41.
Lorimer, Frank, and Frederick Osborn. 1934. *Dynamics of Population: Social and Biological Significance of Changing Birth Rates in the United States.* New York: Macmillan.
Lunde, Anders. 1981. "The Beginning of the Population Association of America." *Population Index* 47, no. 3: 479–484.
Malthus, T. R. 1989 [1798]. *An Essay on the Principle of Population, Volume I,* ed. Patricia James. Cambridge: Cambridge University Press.
Mani, Lata. 1990. "Multiple Mediations: Feminist Scholarship in the Age of Multinational Reception." *Feminist Review* 35 (Summer): 24–41.
Marks, Lara. 2001. *Sexual Chemistry: A History of the Contraceptive Pill.* New Haven: Yale University Press.
Mauldin, W. Parker. 1965. "Fertility Studies: Knowledge, Attitude, and Practice." *Studies in Family Planning* 1, no. 7 (June): 1–10.
Mauldin, W. Parker, Nazli Choucri, Frank Notestein, and Michael Teitelbaum. 1974. "A Report on Bucharest." *Studies in Family Planning* 5, no. 12 (December): 357–395.

May, Elaine Tyler. 1999. *Homeward Bound*. New York: Basic Books.
McCann, Carole. 2009. "Malthusian Men and Demographic Transitions: A Case Study of Hegemonic Masculinity in Mid-Twentieth Century Population Theory." Special issue, *Frontiers* 30, no. 1 (May): 142–171.
———. 1994. *Birth Control Politics in the United States, 1916–1945*. Ithaca: Cornell University Press.
McCann, Carole, and Seung-Kyung Kim. 2013. *The Feminist Theory Reader*. 3rd ed. New York: Routledge.
———. 2010. *The Feminist Theory Reader*. 2nd ed. New York: Routledge.
McCarthy, E. Doyle, and Das Robin. 1985. "American Sociology's Idea of Itself: A Review of the Textbook Literature from the Turn of the Century to the Present." *History of Sociology* 5, no. 2: 21–43.
McClintock, Anne. 1995. *Imperial Leather: Race, Gender and Sexuality in the Colonial Contest*. New York: Routledge.
McIntosh, C. Alison, and Jason Finkle. 1995. "The Cairo Conference on Population and Development: A New Paradigm." *Population and Development Review* 21, no. 2: 223–260.
McNicoll, Geoffrey. 1992. "The Agenda of Population Studies: A Commentary and Complaint." *Population and Development Review* 18, no. 3: 399–420.
Messer-Davidow, Ellen, David R. Shumway, and David J. Sylvan. 1993. *Knowledges: Historical and Critical Studies in Disciplinarity*. Charlottesville: University of Virginia Press.
Metcalf, Barbara, and Thomas Metcalf. 2006. *A Concise History of India*. Cambridge: Cambridge University Press.
Mignolo, Walter. 2007. "Introduction: Coloniality of Power and De-Colonial Thinking." *Cultural Studies* 21, no. 2/3 (March/May): 155–167.
———. 2002. "The Geopolitics of Knowledge and the Colonial Difference." *South Atlantic Quarterly* 101, no. 1 (Winter): 57–96.
Moore, Hugh. 1954. *The Population Bomb*. New York: Hugh Moore Fund.
Moya, Paula. 2001. "Chicana Feminism and Postmodernist Theory." *Signs: Journal of Women in Culture and Society* 26, no. 2 (Winter): 441–483.
Mukerjee, Radhakamal. 1946. *Races, Lands and Food*. New York: Dryden.
———. 1941a. "New Approaches to Population." *Social Forces* 20, no. 2: 141–146.
———. 1941b. "Population Theory and Politics." *American Sociological Review* 6, no. 6 (December): 784–793.
———. 1938. *Food Planning for Four Hundred Millions*. London: Macmillan.
———. 1931. "Introduction." In *The Pressure of Population: Its Effects on Rural Economy in Gorakhpur District*, ed. Jai Krishna Mathur, i–vii. Allahabad: Government Press, United Provinces.
Mukerji, Sarat Chandra. 1933. "Birth Control in India." *Birth Control Review* 17 (July): 173–174.
Murphy, Michelle. 2012. *Seizing the Means of Reproduction: Entanglements of Feminism, Health, and Technoscience*. Durham: Duke University Press.
Nair, Rahul. 2011. "The Construction of a 'Population Problem' in Colonial India,

1919–1947." *Journal of Imperial and Commonwealth History* 39, no. 2 (June): 227–247.
———. 2006. "The Discourse on Population in India, 1870–1960." PhD diss., University of Pennsylvania.
Nobles, Melissa. 2000. *Shades of Citizenship: Race and the Census in Modern Politics.* Stanford: Stanford University Press.
Nortman, Dorothy. 1982. "Measuring the Unmet Need for Contraception to Space and Limit Births." *International Family Planning Perspectives* 8, no. 4 (December): 125–134.
Notestein, Frank. 1982. "Demography in the United States: A Partial Account of the Development of the Field." *Population and Development Review* 8, no. 4: 651–687.
———. 1981. "Memories of the Early Years of the Association." *Population Index* 47, no. 3: 484–488.
———. 1978. "John D. Rockefeller 3rd: A Personal Appreciation." *Population and Development Review* 4, no. 3 (September): 501–508.
———. 1971. "Reminiscences: The Role of Foundations, the Population Association of America, Princeton University and the United Nations in Fostering American Interest in Population Problems." *Milbank Memorial Fund Quarterly* 49 (October): 67–84.
———. 1969a. "Frederick Osborn: Demography's Statesman on His Eightieth Spring." *Population Index* 35, no. 4: 367–371.
———. 1969b. "Population Growth and Its Control." In *Overcoming World Hunger*, ed. American Assembly, 9–39. Englewood Cliffs, NJ: Prentice Hall.
———. 1968. "The Population Council and the Demographic Crisis of the Less Developed World." *Demography* 5, no. 2: 553–560.
———. 1959. "Poverty and Population." *Atlantic* 204, no. 5 (November): 84–87.
———. 1952. "The Economics of Population and Food Supplies: I. Economic Problems of Population Change." *Proceedings of the Eighth International Conference of Agricultural Economists,* 13–31.
———. 1951. "The Needs of World Population." *Bulletin of the Atomic Scientists* 7, no. 4 (April): 99–101 and 128.
———. 1950a. "Demographic Work of the United Nations." *Population Index* 16, no. 3 (July): 184–193.
———. 1950b. "The Reduction of Human Fertility as an Aid to Programs of Economic Development in Densely Settled Agrarian Regions." In *Modernization Programs in Relation to Human Resources and Population Problems,* ed. Milbank Memorial Fund, 89–100. New York: Milbank Memorial Fund.
———. 1948. "Summary of the Demographic Background of Problems of Undeveloped Areas." *Milbank Memorial Fund Quarterly* 26, no. 3: 249–255.
———. 1945a. "International Population Readjustments." *Proceedings of the Academy of Political Science* 21, no. 2 (January): 94–102.
———. 1945b. "Population—The Long View." In *Food for the World*, ed. Theodore Schultz, 36–57. Chicago: University of Chicago Press.

———. 1944. "Problems of Policy in Relation to Areas of Heavy Population Pressure." *Milbank Memorial Fund Quarterly* 22, no. 4 (October): 424–444.
———. 1943. "Some Implications of Population Changes for Post-War Europe." *Proceedings of the American Philosophical Society* 87, no. 2: 165–174.
———. 1939a. "Intrinsic Factors in Population Growth." *Proceedings of the American Philosophical Society* 80, no. 4: 499–511.
———. 1939b. "Some Implications of Current Demographic Trends for Birth Control and Eugenics." *Journal of Heredity* 30, no. 3: 121–126.
Notestein, Frank, and Clyde Kiser. 1935. "Factors Affecting Variations in Human Fertility." *Social Forces* 14, no. 1 (October): 32–41.
O'Connor, Alice. 2001. *Poverty Knowledge: Social Science, Social Policy, and the Poor in Twentieth-Century U.S. History*. Princeton: Princeton University Press.
Ó Gráda, Cormac. 2009. *Famine*. Princeton: Princeton University Press.
Omi, Michael. 1997. "Racial Identity and the State: The Dilemmas of Classification." *Law and Inequality: A Journal of Theory and Practice* 15, no. 1: 7–23.
Osborn, Frederick. 1968. *The Future of Human Heredity: An Introduction to Eugenics in Modern Society*. New York: Weybright and Talley.
———. 1958. *Population: An International Dilemma*. New York: Population Council.
———. 1940. *Preface to Eugenics*. New York: Harper's and Brothers.
———. 1937. "Development of a Eugenic Philosophy." *American Sociological Review* 2, no. 3 (June): 389–397.
Oudshoorn, Nelly. 1994. *Beyond the Natural Body: An Archeology of Sex Hormones*. London: Routledge.
Parry, Manon. 2013. *Broadcasting Birth Control: Mass Media and Family Planning*. New Brunswick: Rutgers University Press.
Pateman, Carole. 1988. *The Sexual Contract*. Palo Alto: Stanford University Press.
Pearl, Raymond. 1934a. "Contraception and Fertility in 4945 Married Women." *Human Biology* 6 (May): 355–401.
———. 1934b. "Second Progress Report on Family Limitation." *Milbank Memorial Fund Quarterly* 11 (July): 248–269.
———. 1933. "Factors in Human Fertility and Their Statistical Evaluation." *The Lancet* 222, no. 5741 (September 9): 607–611.
———. 1932. "Contraception and Fertility in 2,000 Women." *Human Biology* 4, no. 3: 363–407.
Peiss, Kathy. 1986. *Cheap Amusements: Working Women and Leisure in Turn-of-the-Century New York*. Philadelphia: Temple University Press.
Petchesky, Rosalind. 1985. *Abortion and Woman's Choice*. Boston: Northeastern University Press.
Piotrow, Phyllis. 1973. *World Population Crisis: The United States Response*. New York: Praeger.
Platt, Jennifer. 1996. *A History of Sociological Research Methods in America, 1920–1960*. New York: Cambridge University Press.
Poovey, Mary. 1998. *A History of the Modern Fact*. Chicago: University of Chicago Press.

———. 1995. *Making a Social Body: British Cultural Formation, 1830–1964*. Cambridge: Cambridge University Press.

———. 1993. "Figures of Arithmetic, Figures of Speech: The Discourse of Statistics in the 1830s." *Critical Inquiry* 19, no. 2 (Winter): 256–276.

Population Council. 1978. *The Population Council: A Chronicle of the First Twenty-Five Years, 1952–1977*. New York: Population Council.

Porter, Theodore. 2004. *Karl Pearson: The Scientific Life in a Statistical Age*. Princeton: Princeton University Press.

———. 1995. *Trust in Numbers: The Pursuit of Objectivity in Science and Public Life*. Princeton: Princeton University Press.

———. 1986. *The Rise of Statistical Thinking, 1820–1900*. Princeton: Princeton University Press.

Prakash, Gyan. 1999. *Another Reason: Science and the Imagination of Modern India*. Princeton: Princeton University Press.

Prashad, Vijay. 2007. *The Darker Nations: A People's History of the Third World*. New York: New Press.

Presser, Harriet. 1997. "Demography, Feminism, and the Science-Policy Nexus." *Population and Development Review* 23, no. 2: 295–331.

Quijano, Aníbal. 2007. "Coloniality and Modernity/Rationality." *Cultural Studies* 21, no. 2/3 (March/May): 168–178.

Ramsden, Edmund. 2008. "Eugenics from the New Deal to the Great Society: Genetics, Demography and Population Quality." *Studies in the History and Philosophy of the Biological and Biomedical Sciences* 39: 391–406.

———. 2004. "Frank Notestein, Frederick Osborn, and the Development of Demography in the United States." *Princeton University Library Chronicle* 54, no. 2: 282–316.

———. 2003. "Social Demography and Eugenics in the Interwar United States," *Population and Development Review* 49, no. 4: 547–593.

———. 2002. "Carving up Population Science: Eugenics, Demography and the Controversy over the 'Biological Law' of Population Growth." *Social Studies of Science* 32, no. 5/6: 857–899.

———. 2001. "Between Quality and Quantity: The Population Council and the Politics of 'Science-making' in Eugenics and Demography, 1952–1965." Rockefeller Archives Center, Sleepy Hollow, NY.

Ramusack, Barbara. 2006. "Authority and Ambivalence: Medical Women and Birth Control in India." In *Reproductive Health in India: History, Politics, Controversies*, ed. Sarah Hodges, 51–84. New Delhi: Orient Longman.

———. 1989. "Embattled Advocates: The Debate over Birth Control in India." *Journal of Women's History* 1, no. 2 (Fall): 34–64.

Rao, Mohan. 2004. *From Population Control to Reproductive Health: Malthusian Arithmetic*. London: Zed Books.

Rau, Dhanvanthi Rama. 1977. *An Inheritance: The Memoirs of Dhanvanthi Rama Rau*. New York: Harper and Row.

Ray, Raka. 2006. "Is the Revolution Missing or Are We Looking in the Wrong Places?" *Social Problems* 53, no. 4: 459–465.
Reed, James. 1983. *The Birth Control Movement in American Society.* Princeton: Princeton University Press.
———. 1978. *From Private Vice to Public Virtue: The Birth Control Movement and American Society since 1830.* New York: Basic Books.
Riedmann, Agnes. 1993. *Science That Colonizes: A Critique of Fertility Studies in Africa.* Philadelphia: Temple University Press.
Riley, Nancy E., and James McCarthy. 2003. *Demography in the Age of the Post-modern.* Cambridge: Cambridge University Press.
Robbins, John. 1973. "Unmet Needs in Family Planning: A World Survey." *Family Planning Perspectives* 5, no. 4 (Autumn): 232–236.
Rockefeller, John D., III. 1978 [1974]. "Population Growth: The Role of the Developed World." Reprinted in *Population and Development Review* 4, no. 3 (September): 509–516.
Rose, Nikolas. 2007. *The Politics of Life Itself: Biomedicine, Power and Subjectivity in the Twenty-First Century.* Princeton: Princeton University Press.
———. 1991. "Governing by Numbers: Figuring Our Democracy." *Accounting Organizations and Society* 16: 673–692.
Rose, Nikolas, and Peter Miller. 1992. "Political Power beyond the State: Problematics of Government." *British Journal of Sociology* 43, no. 2: 173–205.
Rotman, Brian. 2000. *Mathematics as Sign: Writing, Imagining, Counting.* Palo Alto: Stanford University Press.
Ryder, Norman. 1984. "Frank Wallace Notestein (1902–1983)." *Population Studies* 38, no. 1: 5–20.
Sanger, Margaret. 1939. *An Autobiography.* London: Victor Gollancz.
———. 1928. *Motherhood in Bondage.* New York: Brentano's Publishers.
———, ed. 1927. *Proceedings of the World Population Conference, 1927.* London: Edward Arnold.
Sanger, Margaret, and Hannah Stone. 1931. *The Practice of Contraception.* Baltimore: Williams and Wilkins.
Schoen, Johanna. 2005. *Coercion and Choice: Birth Control, Sterilization, and Abortion in Public Health and Welfare.* Chapel Hill: University of North Carolina Press.
Schweber, Libby. 2007. *Disciplining Statistics: Demography and Vital Statistics in France and England, 1830–1885.* Durham: Duke University Press.
"Science: Eugenics for Democracy." 1940. *Time Magazine* 36, no. 11 (September 9): 34.
Sen, Amartya. 1994. "Population: Delusion and Reality." *The New York Review of Books* 41, no. 15 (September 22): 62–71.
Sen, Gita, Adrienne Germain, and Lincoln Chen. 1994. *Population Policy Reconsidered: Health, Empowerment, Rights.* Cambridge: Harvard University Press.
Sharpless, John. 1995. "World Population Growth, Family Planning and American Foreign Policy." *Journal of Policy History* 7, no. 1: 72–102.

Silliman, Jael, and Ynestra King, eds. 1999. *Dangerous Intersections: Feminist Perspectives on Population, Environment, and Development.* Cambridge, MA: South End Press.
Silverberg, Helene. 1998. *Gender and American Social Science: The Formative Years.* Princeton: Princeton University Press.
Simmons, Christina. 2009. *Making Marriage Modern: Women's Sexuality from the Progressive Era to World War II.* New York: Oxford University Press.
Sinha, Mrinalini. 2006. *Specters of Mother India: The Global Restructuring of an Empire.* Durham: Duke University Press.
———. 2006. "Gender and Nation." In *Women's History in Global Perspective,* ed. Bonnie Smith, 229–274. Bloomington: Indiana University Press.
Smith, Dorothy. 2006. *Institutional Ethnography as Practice.* Lanham, MD: Rowman and Littlefield.
———. 2005. *Institutional Ethnography: A Sociology for People.* Walnut Creek, CA: AltaMira Press.
———. 1990. *The Conceptual Practices of Power.* Boston: Northeastern University Press.
———. 1987. *The Everyday World as Problematic: A Feminist Sociology.* Boston: Northeastern University Press.
Somers, Margaret. 1996. "Where Is Sociology after the Historical Turn? Knowledge Cultures, Narrativity, and Historical Epistemologies." In *The Historical Turn in the Human* Sciences, ed. Terrence J. McDonald, 53–89. Ann Arbor: University of Michigan Press.
Spivak, Gayatri. 1992. "Acting Bits/Identity Talk." *Critical Inquiry* 18, no. 4: 770–803.
Stacey, Judith, and Barrie Thorne. 1985. "The Missing Feminist Revolution in Sociology." *Social Problems* 32, no. 4: 301–316.
Stern, Alexandra. 2005. *Eugenic Nation: Faults and Frontiers of Better Breeding in Modern America.* Berkeley: University of California Press.
Stix, Regine, and Frank Notestein. 1940. *Controlled Fertility: An Evaluation of Clinic Service.* Baltimore: Williams and Wilkins.
———. 1935. "The Effectiveness of Birth Control: A Second Study of Contraceptive Practice in a Selected Group of New York Women." *Milbank Memorial Fund Quarterly* 13, no. 2: 162–178.
———. 1934. "The Effectiveness of Birth Control: A Study of Contraceptive Practice in a Selected Group of New York Women." *Milbank Memorial Fund Quarterly* 12, no. 1: 57–68.
Stoler, Ann Laura. 1995. *Race and the Education of Desire: Foucault's History of Sexuality and the Colonial Order of Things.* Durham: Duke University Press.
Stone, Abraham. 1953. "Fertility Problems in India." *Fertility and Sterility* 4, no. 3 (May/June): 210–217.
Stone, Hannah. 1933. "Maternal Health and Contraception: Medical Data." *Medical Journal and Record,* April 19 and May 3.
Subramaniam, Banu. 2014. *Ghost Stories for Darwin: The Science of Variation and the Politics of Diversity.* Urbana: University of Illinois Press.

Suitters, Beryl. 1973. *Be Brave and Angry: Chronicles of the International Planned Parenthood Federation*. London: International Planned Parenthood Federation.

Symonds, Richard, and Michael Carder. 1973. *The United Nations and the Population Question*. New York: McGraw-Hill.

Szreter, Simon. 1996. *Fertility, Class and Gender in Britain, 1860–1940*. Cambridge: Cambridge University Press.

———. 1993. "The Idea of Demographic Transition and the Study of Fertility: A Critical Intellectual History." *Population and Development Review* 19, no. 4: 659–701.

Szreter, Simon, Hania Sholkamy, and A. Dharmalingam. 2004. *Categories and Contexts: Anthropological and Historical Studies in Critical Demography*. Oxford: Oxford University Press.

Taeuber, Irene. 1958. *The Population of Japan*. Princeton: Princeton University Press.

———. 1946. "Population Studies in the United States." *Population Index* 12: 254–269.

Takeshita, Chikako. 2011. *The Global Biopolitics of the IUD: How Science Constructs Contraceptive Users and Women's Bodies*. Cambridge: MIT Press.

Taylor, Charles. 2004. *Modern Social Imaginaries*. Durham, NC: Duke University Press.

Thakur, Manish. 2012. "Radhakamal Mukerjee and the Quest for an Indian Sociology." *Sociological Bulletin* 61, no. 6 (January/April): 89–108.

Thompson, S. Warren. 1929. *Danger Spots in World Population*. New York: Alfred A. Knopf.

United Nations. 2014 [1994]. "Programme of Action Adopted at the International Conference on Population Development: 20th Anniversary Edition." New York: United Nations.

———. 1952. "Accuracy Tests for Census Age Distribution in Five-Year and Ten-Year Groups." *Population Bulletin*, no. 2, 69–79. New York: United Nations.

———. 1951. *Population Bulletin*, no. 1. New York: United Nations.

Vance, Rupert. 1959. "The Development and Status of American Demography." In *The Study of Population: An Inventory and Appraisal*, ed. Philip Hauser and Otis Dudley Duncan, 286–313. Chicago: University of Chicago Press.

Van de Tak, Jean. 1991. *Demographic Destinies: Interviews with Presidents and Secretary-Treasurers of the Population Association of America, Volumes 1 and 2*. Washington, DC: Population Association of America.

Wanzo, Rebecca. 2009. *The Suffering Will Not Be Televised: African American Women and Sentimental Political Storytelling*. New York: SUNY Press.

Watkins, Elizabeth, 1998. *On the Pill: A Social History of Oral Contraceptives, 1950–1970*. Baltimore: Johns Hopkins University Press.

Watkins, Susan Cott. 1993. "If All We Knew about Women Was What We Read in *Demography*, What Would We Know?" *Demography* 30, no. 4: 551–577.

Wattal, Pyare Krishan. 1916. *The Population Problem in India: A Census Study*. Bombay: Bennett, Coleman.

Weinbaum, Alys Eve, Lynn Thomas, Priti Ramamurthy, Uta Poiger, Madeline Yue Dong, and Tani Barlow. 2008. *Modern Girl around the World: Consumption, Modernity, and Globalization.* Durham: Duke University Press.
Westoff, Charles. 1994. "What's the World's Priority Task? Finally, Control Population." *New York Times,* February 6.
———. 1992. "Measuring the Unmet Need for Contraception: Comment on Bongaarts." *Population and Development Review* 18, no. 1: 123–125.
———. 1988a. "Is the KAP-Gap Real?" *Population and Development Review* 14, no. 2 (June): 225–232.
———. 1988b. "The Potential Demand for Family Planning: A New Measure of Unmet Need and Estimates for Five Latin American Countries." *International Family Planning Perspectives* 14, no. 2 (June): 45–53.
———. 1981. "Unwanted Fertility in Six Developing Countries." *International Family Planning Perspectives* 7, no. 2 (June): 43–52.
———. 1978. "The Unmet Need for Birth Control in Five Asian Countries." *International Family Planning Perspectives* 4, no. 1 (Spring): 9–18.
Westoff, Charles, and Anne Pebley. 1981. "Alternative Measures of Unmet Need for Family Planning in Developing Countries." *International Family Planning Perspectives* 7, no. 4 (December): 126–136.
Whelpton, Pascal. 1938. *Needed Population Research.* Lancaster, PA: Science Press.
Whelpton, Pascal, and Clyde Kiser. 1945. "Trends, Determinants, and Control of Human Fertility." *Annals of the American Academy of Political and Social Science* 237 (January): 112–122.
Wilmoth, John, and Patrick Ball. 1992. "The Population Debate in American Popular Magazines, 1946–1990." *Population and Development Review* 18, no. 4 (December): 631–668.
Wilson, Chris, Jim Oeppen, and Mike Pardoe. 1988. "What Is Natural Fertility? The Modelling of a Concept." *Population Index* 54, no. 1 (Spring): 4–20.
Winch, Donald. 1987. *Malthus.* Oxford: Oxford University Press.
"Women's Declaration on Population Policy." 1994. In *Population Policy Reconsidered: Health, Empowerment, Rights,* ed. Gita Sen, Adrienne Germain, and Lincoln Chen, 31–34. Cambridge: Harvard University Press.
World Population Conference. 1975. "World Population Plan of Action." *Population and Development Review* 1, no. 1 (September): 163–181.
Young, Robert, J. C. 2001. *Postcolonialism: An Historical Introduction.* Oxford: Blackwell.
———. 1995. *Colonial Desire: Hybridity in Theory, Culture, and Race.* London: Routledge.
Young, Ruth. 1935. "Some Aspects of Birth Control in India." *Marriage Hygiene,* series 1, vol. 2 (August): 37–42.
Yuval-Davis, Nira. 1997. *Gender and Nation.* Thousand Oaks, CA: Sage.
Zachariah, Benjamin. 2001. "Uses of Scientific Argument: The Case of 'Development' in India, c 1930–1950." *Economic and Political Weekly* 36, no. 39 (September 29): 3689–3702.

Index

C

D

E

G

H

N

O

P

Q

R

Z

Lightning Source UK Ltd.
Milton Keynes UK
UKHW011323171222
414000UK00003BA/117